student study
ART NOTEBOOK

eighth edition
Inquiry into Life

Sylvia S. Mader

WCB Wm. C. Brown Publishers

Dubuque, IA   Bogotá   Buenos Aires   Caracas   Chicago   Guilford, CT   London
Madrid   Mexico City   Seoul   Singapore   Sydney   Taipei   Tokyo   Toronto

A Times Mirror Company

The credits section for this book begins on page 215 and is considered
an extension of the copyright page.

Cover photo credit: © Kevin Schafer

A Times Mirror Company

ISBN 0–697–25191–8

Printed in the United States of America by Wm. C. Brown Communications, Inc.,
2460 Kerper Boulevard, Dubuque, IA 52001

This Student Study Art Notebook is an ancillary to assist students in note taking during lectures. On each page, there are one, two, or sometimes three figures reproduced from the textbook. Each figure also corresponds to one of the 250 transparencies available to instructors who adopt this textbook.

The intention is to place the transparency art in front of students (via the notebook) as the instructor uses the overhead during lectures. The advantage to the student is that he/she will be able to see all labels clearly, and take meaningful notes without having to make hurried sketches of the transparency figure.

The pages of the Art Notebook are perforated and three-hole punched, so they can be removed and placed in a personal binder for specific study and review, or to create space for additional notes.

DIRECTORY OF NOTEBOOK FIGURES

TO ACCOMPANY

SYLVIA S. MADER *INQUIRY INTO LIFE, 8E.*

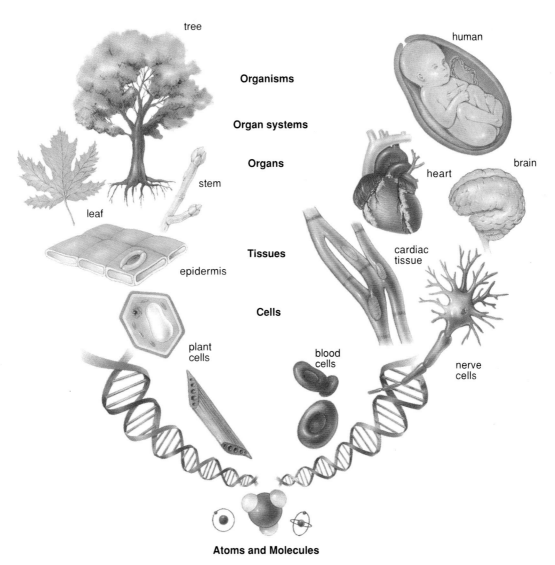

tree

Organisms

human

Organ systems

Organs

stem

heart

brain

leaf

Tissues

epidermis

cardiac tissue

Cells

plant cells

blood cells

nerve cells

Atoms and Molecules

Levels of Organization
Figure 1.2

Kingdoms of Life	Representative Organisms	Organization	Type of Nutrition	Representative Organisms
Monera		Small, simple, single cell (sometimes chains or mats)	Absorb food (some photosynthesize)	Bacteria including cyanobacteria
Protista		Complex single cell (sometimes chains or colonies)	Absorb, photosynthesize, or ingest food	Protozoans, algae, water molds
Fungi		Multicellular filamentous form with specialized complex cells	Absorb food	Molds and mushrooms
Plantae		Multicellular form with specialized complex cells	Photosynthesize food	Mosses, ferns, pine trees, woody and non-woody flowering plants
Animalia		Multicellular form with specialized complex cells	Ingest food	Worms, sponges, insects, fish, reptiles, amphibians, birds, and mammals

Classification of Organisms
Figure 1.7

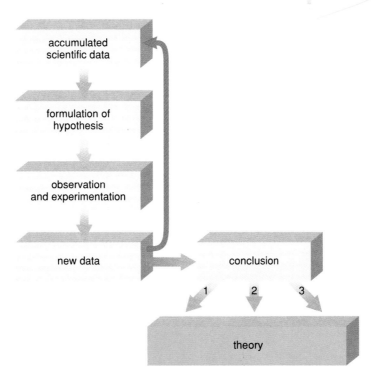

Flow Diagram for the Scientific Method
Figure 1.8

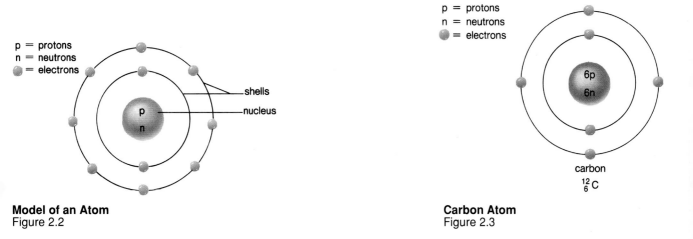

Model of an Atom
Figure 2.2

Carbon Atom
Figure 2.3

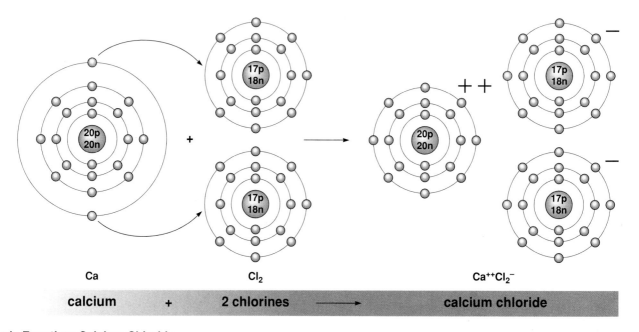

Ca   Cl₂   Ca⁺⁺Cl₂⁻

calcium   **+**   **2 chlorines**   ⟶   **calcium chloride**

Ionic Reaction: Calcium Chloride
Figure 2.5

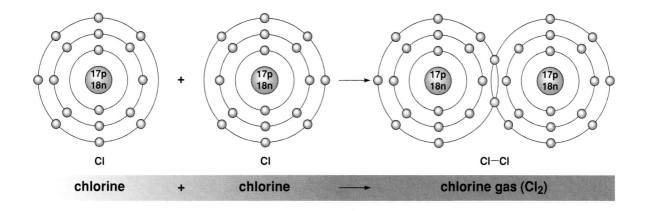

CI + CI → CI—CI

chlorine + chlorine ⟶ chlorine gas (Cl_2)

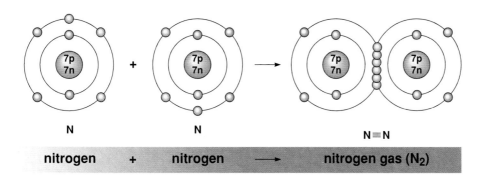

N + N → N≡N

nitrogen + nitrogen ⟶ nitrogen gas (N_2)

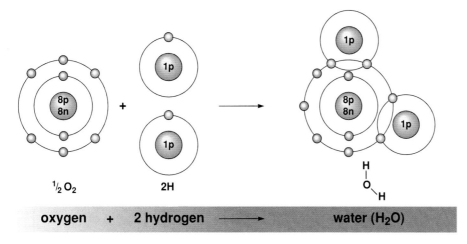

$^1/_2 O_2$ + 2H

H
O
H

oxygen + 2 hydrogen ⟶ water (H_2O)

Covalent Reactions
Figure 2.6

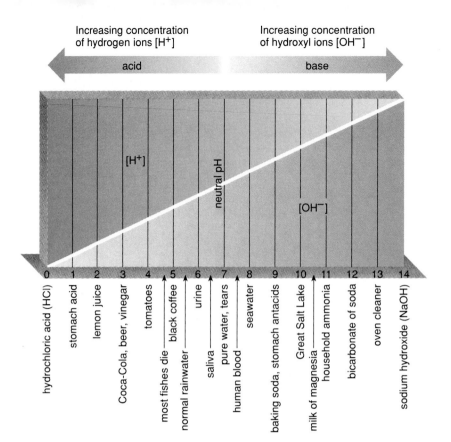

Increasing concentration
of hydrogen ions [H⁺]

Increasing concentration
of hydroxyl ions [OH⁻]

acid

base

$[H^+]$

neutral pH

$[OH^-]$

0 1 2 3 4 5 6 7 8 9 10 11 12 13 14

hydrochloric acid (HCl)
stomach acid
lemon juice
Coca-Cola, beer, vinegar
tomatoes
most fishes die
black coffee
normal rainwater
urine
saliva
pure water, tears
human blood
seawater
baking soda, stomach antacids
Great Salt Lake
milk of magnesia
household ammonia
bicarbonate of soda
oven cleaner
sodium hydroxide (NaOH)

The pH Scale
Figure 2.13

CH_2OH

$C_6H_{12}O_6$

CH_2OH

Three Ways to Represent the Structure of Glucose
Figure 2.17

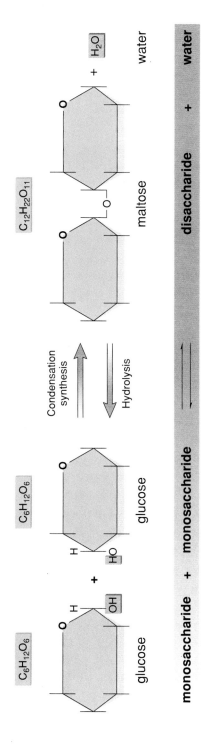

Condensation Synthesis and Hydrolysis of Maltose, a Disaccharide
Figure 2.18

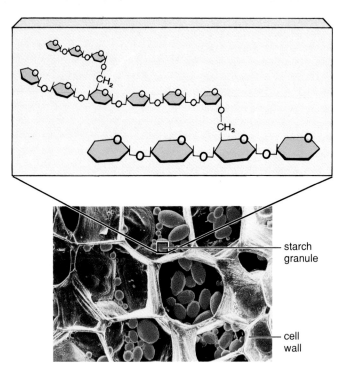

Starch Structure and Function
Figure 2.19

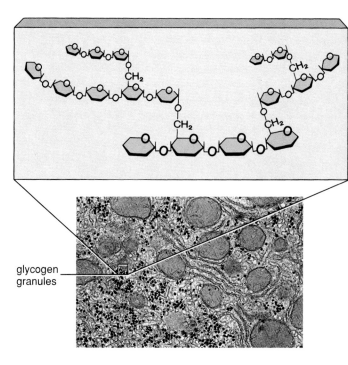

Glycogen Structure and Function
Figure 2.20

Cellulose Structure and Function
Figure 2.21

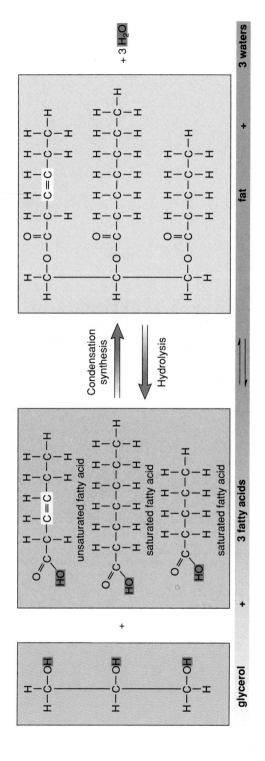

Condensation Synthesis and Hydrolysis of a Fat Molecule
Figure 2.22

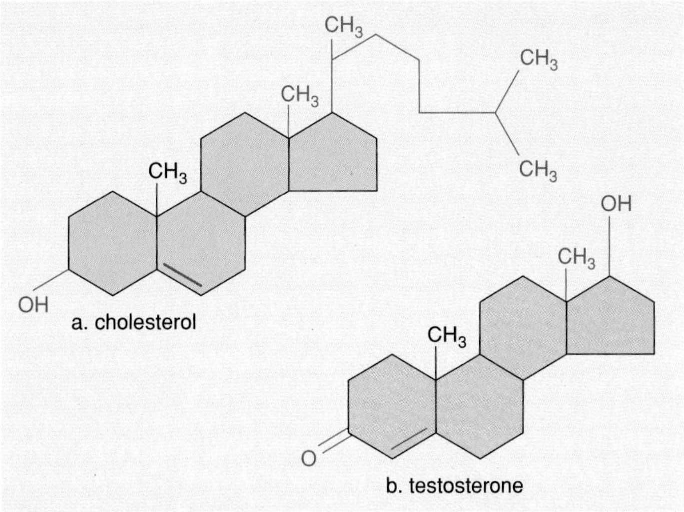

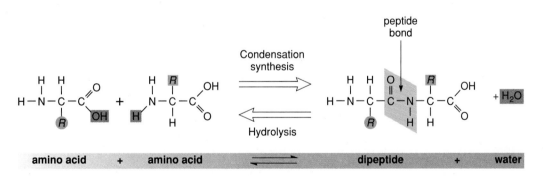

Steroid Diversity
Figure 2.24

Condensation Synthesis and Hydrolysis of a Peptide
Figure 2.26

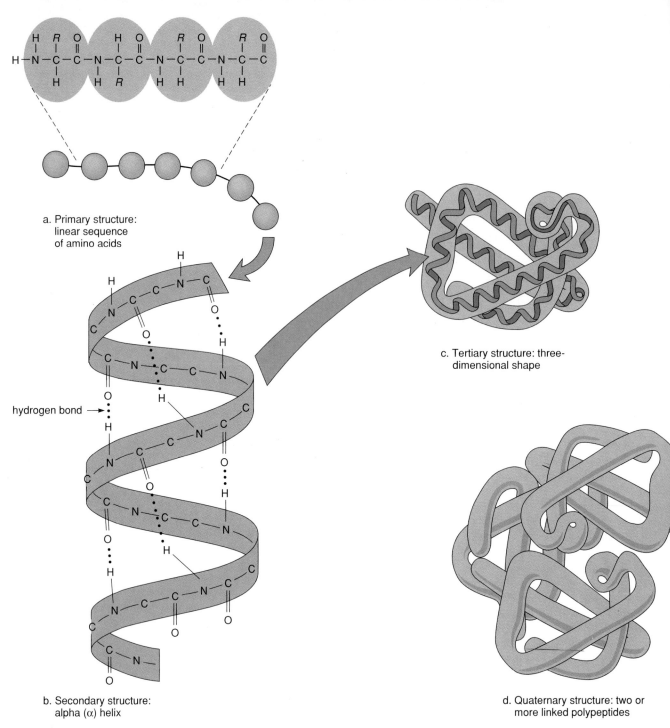

a. Primary structure: linear sequence of amino acids

hydrogen bond →

b. Secondary structure: alpha (α) helix

c. Tertiary structure: three-dimensional shape

d. Quaternary structure: two or more linked polypeptides

Protein Structure
Figure 2.27

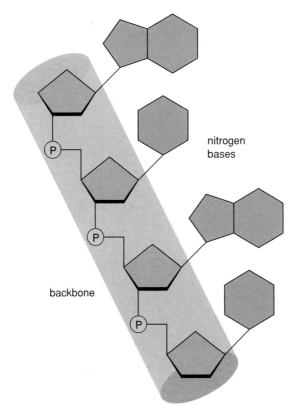

nitrogen
bases

backbone

Generalized Nucleic Acid Strand
Figure 2.29

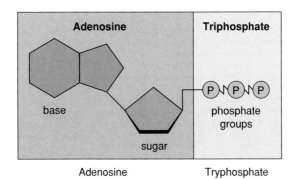

Adenosine   **Triphosphate**

base

phosphate
groups

sugar

Adenosine   Tryphosphate

Structure of ATP
Figure 2.30

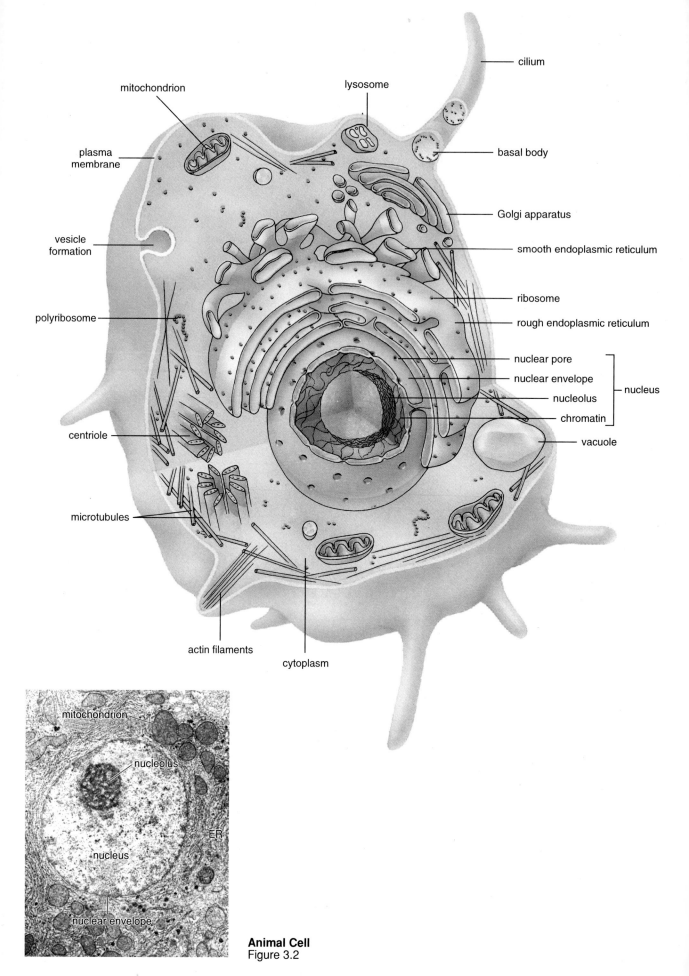

cilium

mitochondrion

lysosome

plasma membrane

basal body

Golgi apparatus

vesicle formation

smooth endoplasmic reticulum

ribosome

rough endoplasmic reticulum

polyribosome

nuclear pore

nuclear envelope

nucleolus

chromatin

nucleus

centriole

vacuole

microtubules

actin filaments

cytoplasm

mitochondrion

nucleolus

ER

nucleus

nuclear envelope

Animal Cell
Figure 3.2

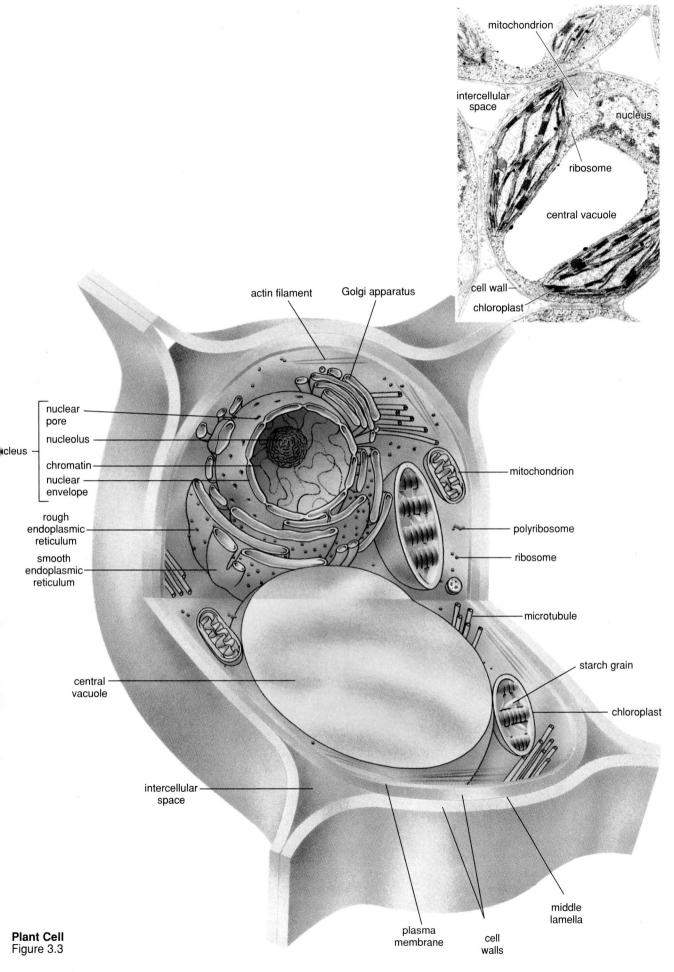

mitochondrion

intercellular space

nucleus

ribosome

central vacuole

cell wall

chloroplast

actin filament

Golgi apparatus

nuclear pore

nucleolus

cleus

chromatin

nuclear envelope

rough endoplasmic reticulum

smooth endoplasmic reticulum

central vacuole

intercellular space

mitochondrion

polyribosome

ribosome

microtubule

starch grain

chloroplast

middle lamella

plasma membrane

cell walls

Plant Cell
Figure 3.3

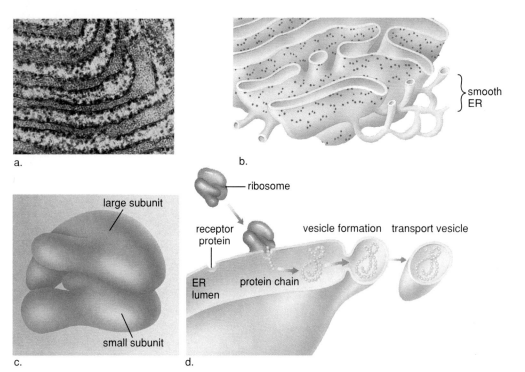

a.

b.

smooth
ER

ribosome

large subunit

receptor
protein

vesicle formation   transport vesicle

ER
lumen

protein chain

small subunit

c.

d.

Rough Endoplasmic Reticulum
Figure 3.5

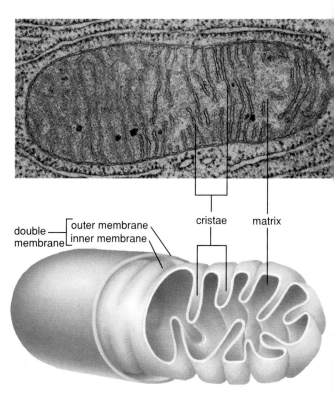

cristae   matrix

double ⌈outer membrane
membrane⌊inner membrane

Mitochondrial Structure
Figure 3.7

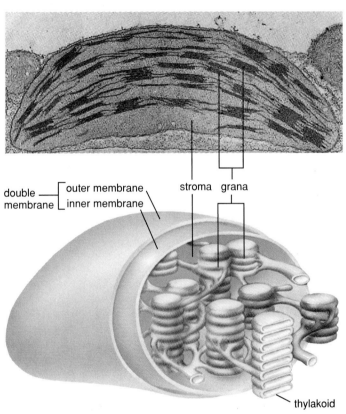

double — ⌈ outer membrane
membrane ⌊ inner membrane

stroma   grana

thylakoid

Chloroplast Structure
Figure 3.8

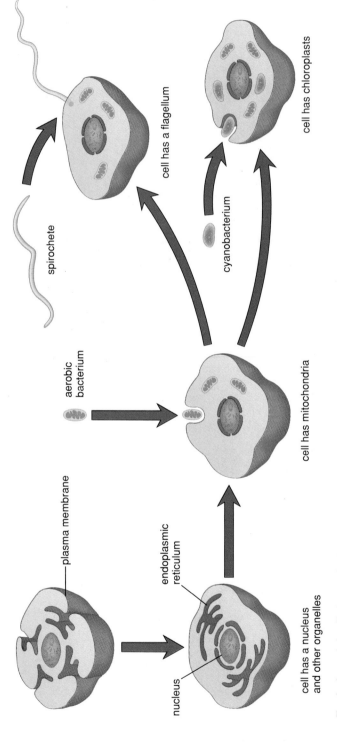

Evolution of the Eukaryotic Cell
Figure 3.9

plasma membrane

endoplasmic reticulum

nucleus

cell has a nucleus and other organelles

aerobic bacterium

cell has mitochondria

spirochete

cell has a flagellum

cyanobacterium

cell has chloroplasts

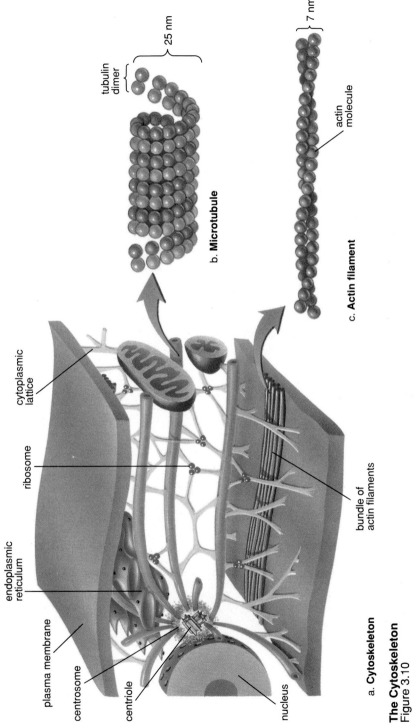

25 nm

tubulin
dimer

b. **Microtubule**

7 nm

actin
molecule

c. **Actin filament**

cytoplasmic
lattice

ribosome

endoplasmic
reticulum

plasma membrane

centrosome

centriole

bundle of
actin filaments

nucleus

a. **Cytoskeleton**

The Cytoskeleton
Figure 3.10

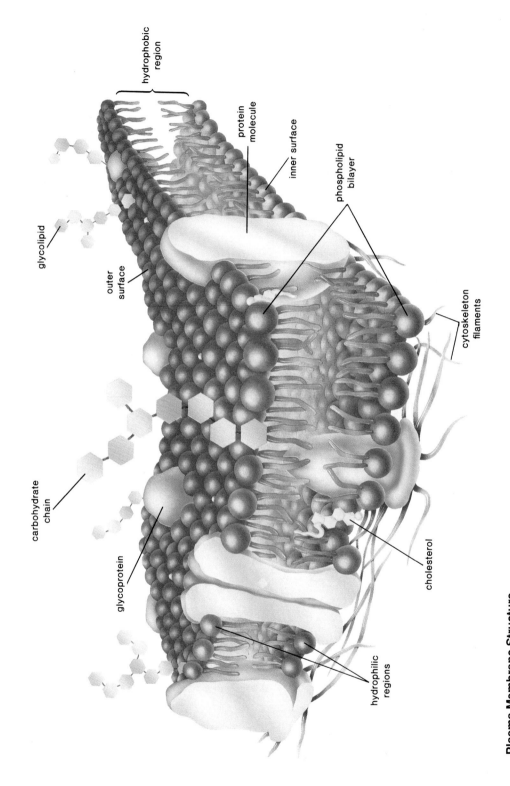

Plasma Membrane Structure
Figure 4.1

Animal Cells

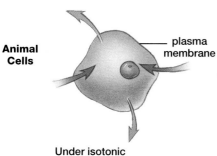

plasma membrane

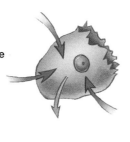

Under isotonic conditions, there is no net movement of water.

In a hypotonic environment, water enters the cell, which may burst (lysis) due to osmotic pressure.

In a hypertonic environment, water leaves the cell, which shrivels (crenation).

Plant Cells

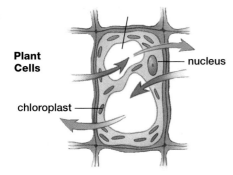

nucleus

chloroplast

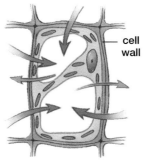

cell wall

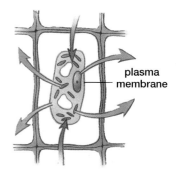

plasma membrane

Under isotonic conditions, there is no net movement of water.

In a hypotonic environment, vacuoles fill with water, turgor pressure develops, and chloroplasts are seen next to the cell wall.

In a hypertonic environment, vacuoles lose water, the cytoplasm shrinks (plasmolysis), and chloroplasts are seen in the center of the cell.

Osmosis in Animal and Plant Cells
Figure 4.7

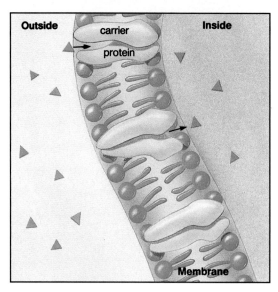

Facilitated Transport
Figure 4.8

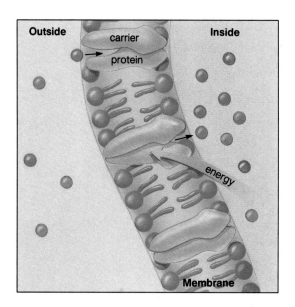

Active Transport
Figure 4.9

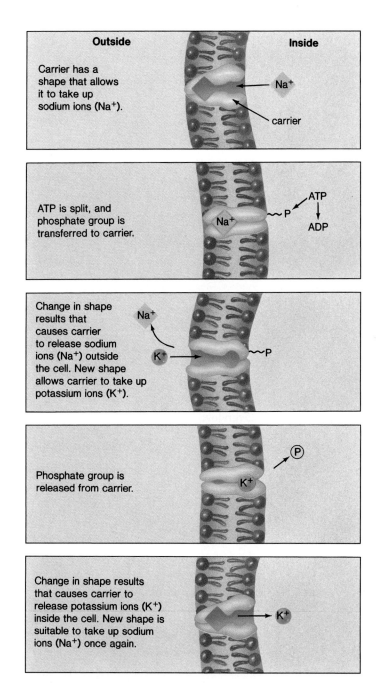

The Sodium-Potassium Pump
Figure 4.10

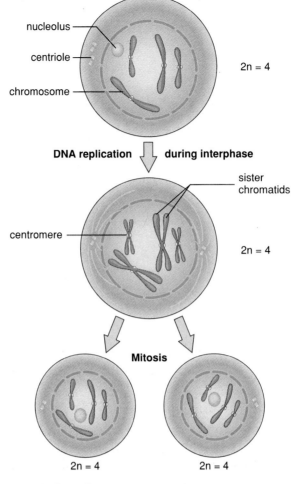

Mitosis Overview
Figure 5.1

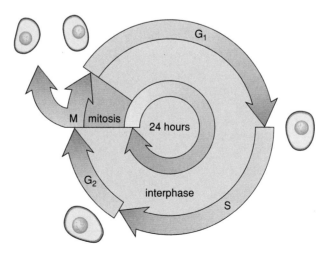

The Cell Cycle
Figure 5.2

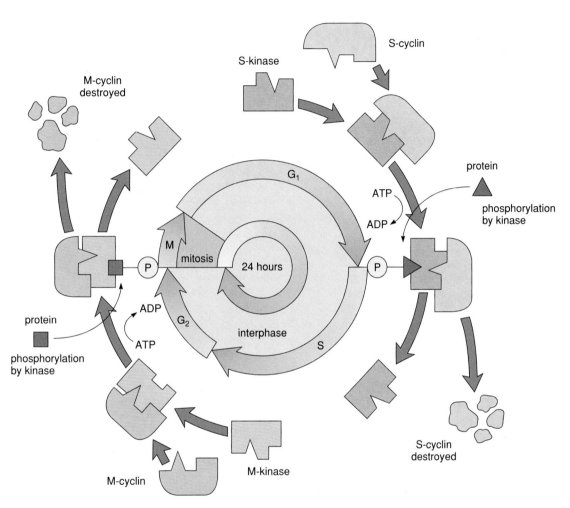

Control of the Cell Cycle
Figure 5.3

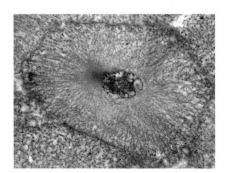

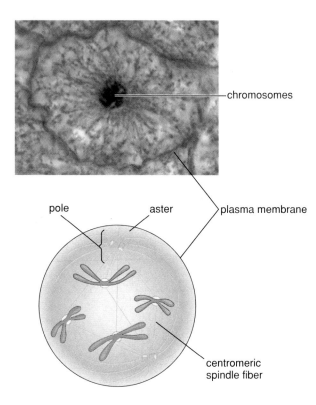

chromosomes

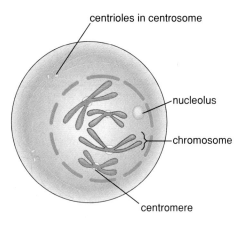

centrioles in centrosome

nucleolus

chromosome

centromere

Prophase

Chromosomes are now distinct. Centrosomes begin moving apart; nuclear envelope is fragmenting and nucleolus will disappear.

pole   aster   plasma membrane

centromeric spindle fiber

Prometaphase

Spindle is in process of forming and centromeres of chromosomes are attaching to centromeric spindle fibers.

Prophase and Prometaphase
Figure 5.4

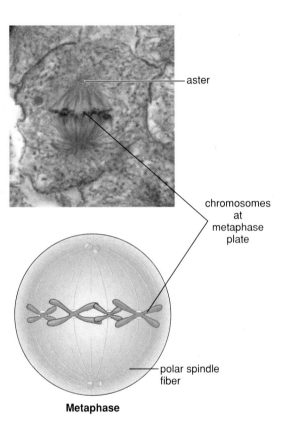

aster

chromosomes
at
metaphase
plate

polar spindle
fiber

Metaphase

Chromosomes (each consisting of
two sister chromatids) are at the
metaphase plate (center of fully
formed spindle).

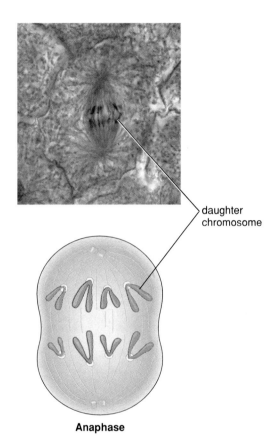

daughter
chromosome

Anaphase

Daughter chromosomes (each
consisting of one chromatid) are at
the poles of the spindle.

Metaphase and Anaphase
Figure 5.5

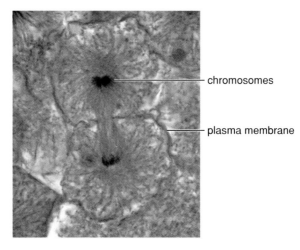

chromosomes

plasma membrane

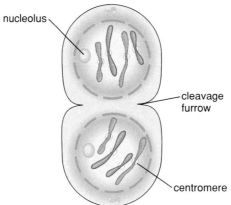

nucleolus

cleavage furrow

centromere

Telophase

Daughter cells are forming as nuclear envelopes and nucleoli appear. Chromosomes will become indistinct chromatin.

Telophase
Figure 5.6

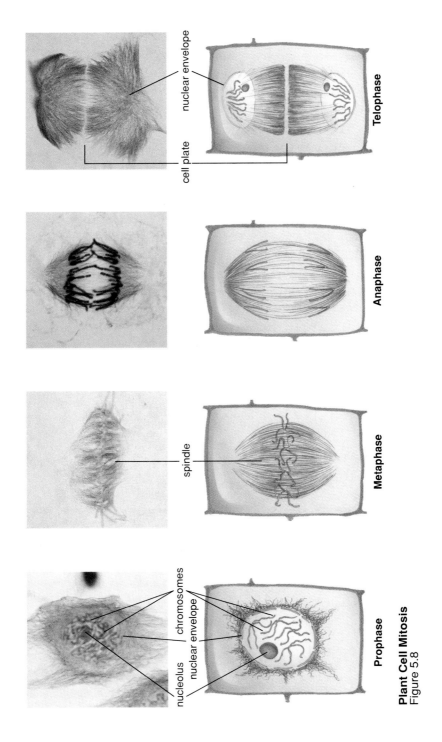

Plant Cell Mitosis
Figure 5.8

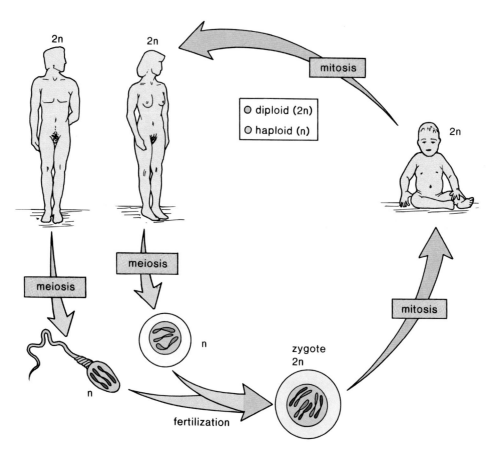

Life Cycle of Humans
Figure 5.9

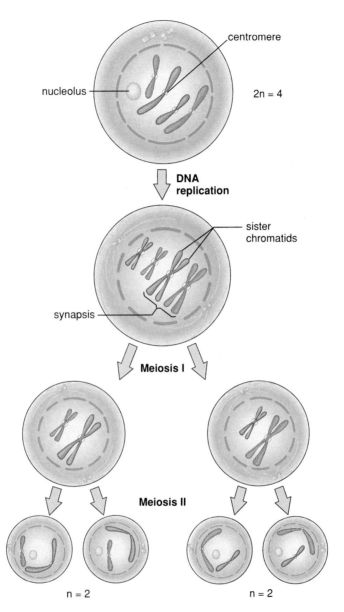

centromere

nucleolus

2n = 4

DNA replication

sister chromatids

synapsis

Meiosis I

Meiosis II

n = 2

n = 2

Meiosis Overview
Figure 5.10

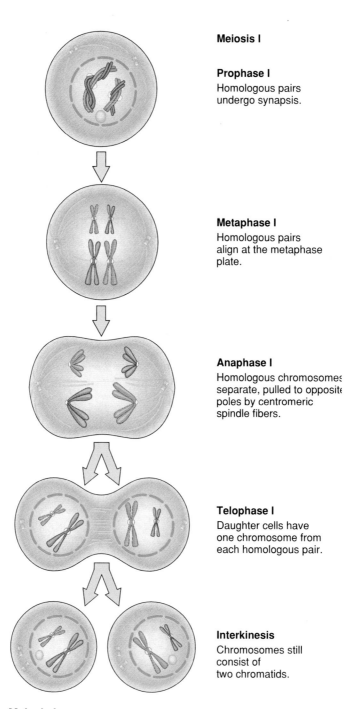

Meiosis I

Prophase I
Homologous pairs
undergo synapsis.

Metaphase I
Homologous pairs
align at the metaphase
plate.

Anaphase I
Homologous chromosomes
separate, pulled to opposite
poles by centromeric
spindle fibers.

Telophase I
Daughter cells have
one chromosome from
each homologous pair.

Interkinesis
Chromosomes still
consist of
two chromatids.

Meiosis I
Figure 5.12

Meiosis II

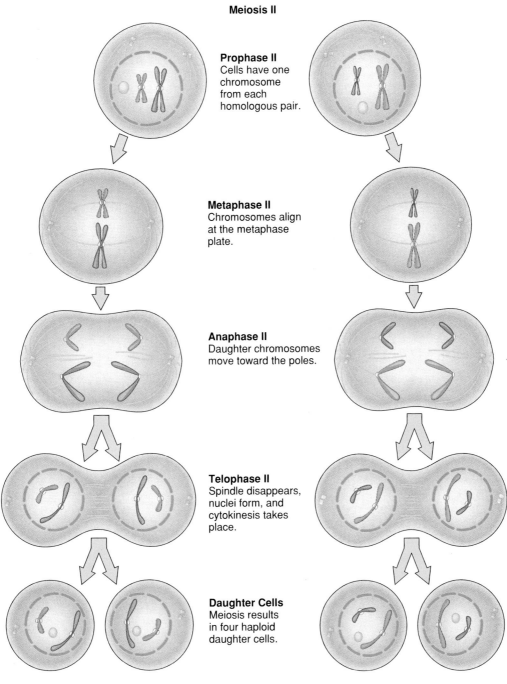

Prophase II
Cells have one chromosome from each homologous pair.

Metaphase II
Chromosomes align at the metaphase plate.

Anaphase II
Daughter chromosomes move toward the poles.

Telophase II
Spindle disappears, nuclei form, and cytokinesis takes place.

Daughter Cells
Meiosis results in four haploid daughter cells.

Meiosis II
Figure 5.13

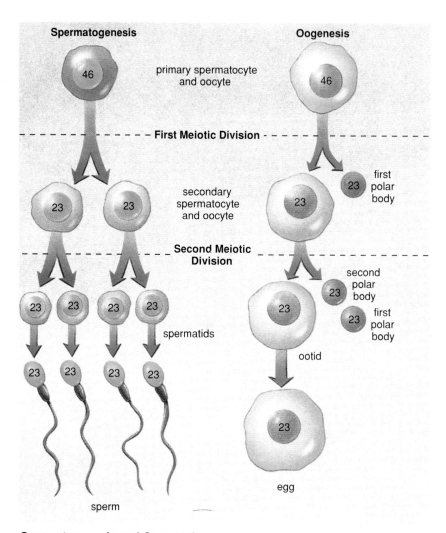

Spermatogenesis and Oogenesis
Figure 5.14

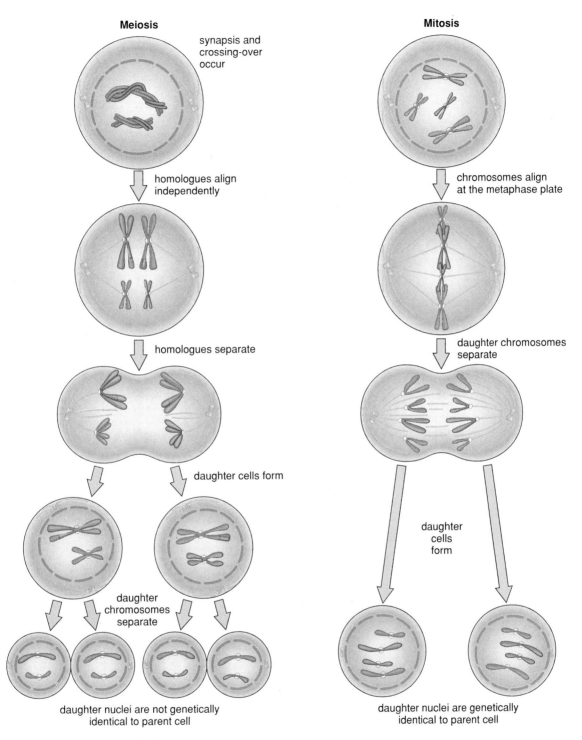

Meiosis

synapsis and crossing-over occur

homologues align independently

homologues separate

daughter cells form

daughter chromosomes separate

daughter nuclei are not genetically identical to parent cell

Mitosis

chromosomes align at the metaphase plate

daughter chromosomes separate

daughter cells form

daughter nuclei are genetically identical to parent cell

Mitosis Compared to Meiosis
Figure 5.16

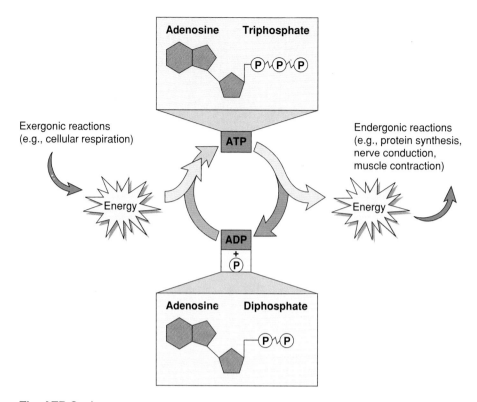

The ATP Cycle
Figure 6.3

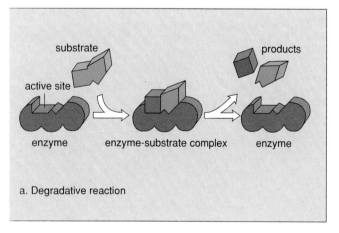

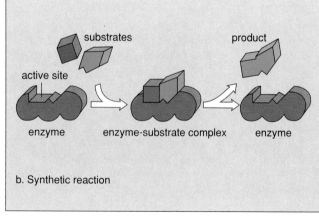

Enzymatic Action
Figure 6.5

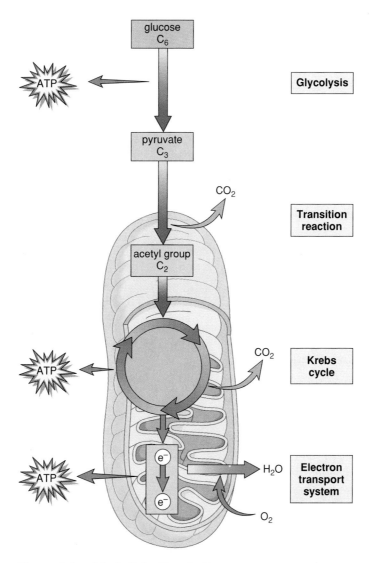

Steps of Aerobic Cellular Respiration
Figure 7.2

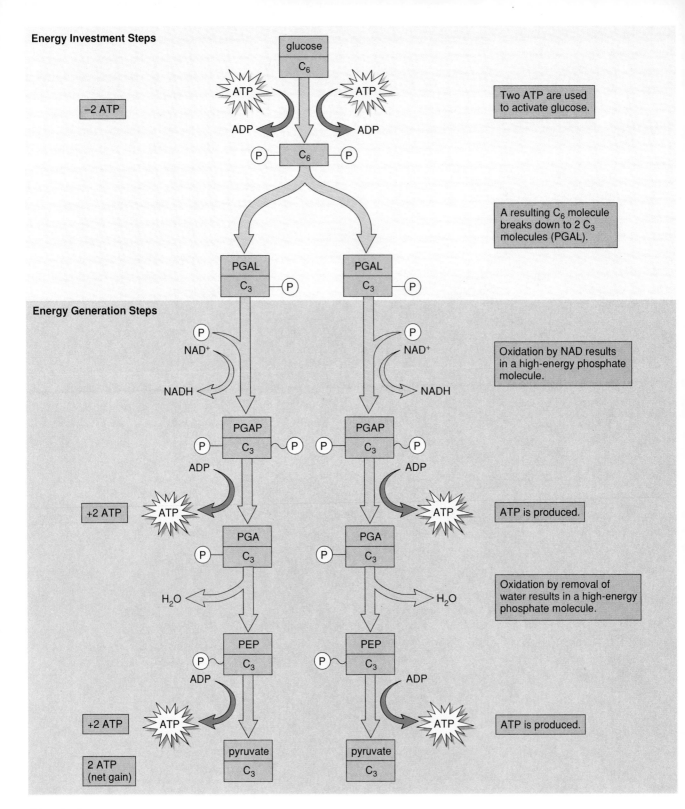

Energy Investment Steps

−2 ATP

Two ATP are used to activate glucose.

glucose C$_6$

ATP → ADP

ATP → ADP

P — C$_6$ — P

A resulting C$_6$ molecule breaks down to 2 C$_3$ molecules (PGAL).

PGAL C$_3$ — P

PGAL C$_3$ — P

Energy Generation Steps

P
NAD$^+$
NADH

P
NAD$^+$
NADH

Oxidation by NAD results in a high-energy phosphate molecule.

PGAP C$_3$ — P

PGAP C$_3$ — P

ADP

ADP

+2 ATP

ATP

ATP

ATP is produced.

PGA C$_3$

PGA C$_3$

H$_2$O

H$_2$O

Oxidation by removal of water results in a high-energy phosphate molecule.

PEP C$_3$

PEP C$_3$

ADP

ADP

+2 ATP

ATP

ATP

ATP is produced.

2 ATP (net gain)

pyruvate C$_3$

pyruvate C$_3$

Glycolysis
Figure 7.4

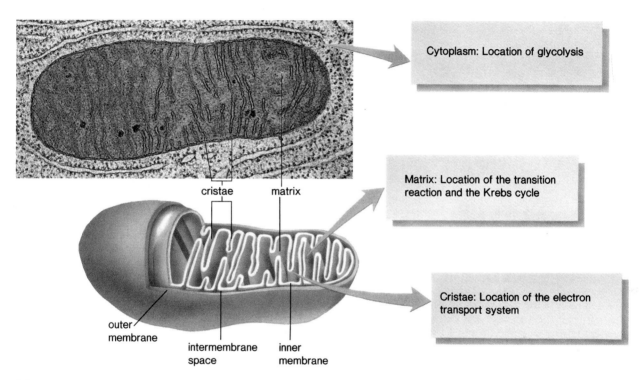

cristae · matrix

Cytoplasm: Location of glycolysis

Matrix: Location of the transition reaction and the Krebs cycle

Cristae: Location of the electron transport system

outer membrane

intermembrane space

inner membrane

Mitochondrion Structure and Function
Figure 7.5

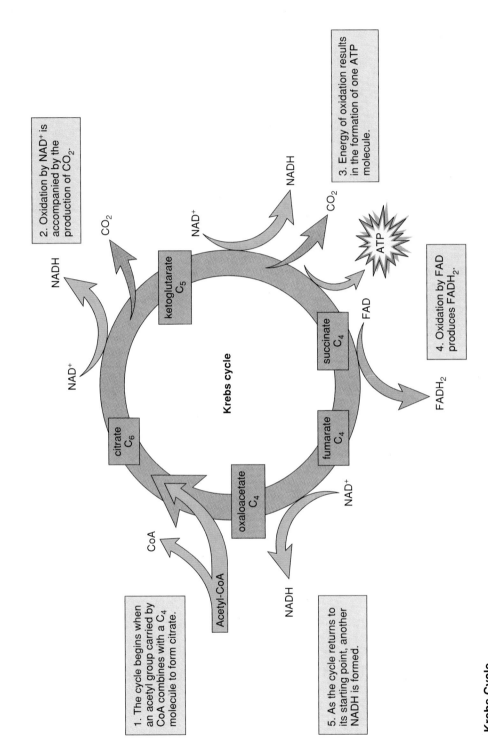

1. The cycle begins when an acetyl group carried by CoA combines with a C_4 molecule to form citrate.

2. Oxidation by NAD^+ is accompanied by the production of CO_2.

3. Energy of oxidation results in the formation of one ATP molecule.

4. Oxidation by FAD produces $FADH_2$.

5. As the cycle returns to its starting point, another NADH is formed.

Krebs cycle

citrate C_6

ketoglutarate C_5

succinate C_4

fumarate C_4

oxaloacetate C_4

Acetyl-CoA

CoA

NADH

NAD^+

CO_2

NAD^+

NADH

CO_2

FAD

$FADH_2$

ATP

NAD^+

NADH

Krebs Cycle
Figure 7.6

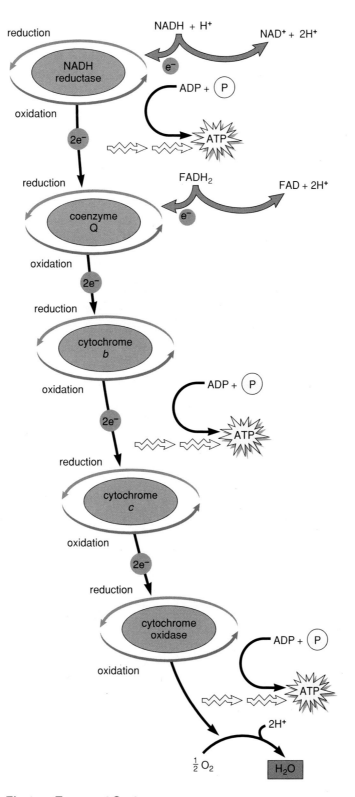

Electron Transport System
Figure 7.7

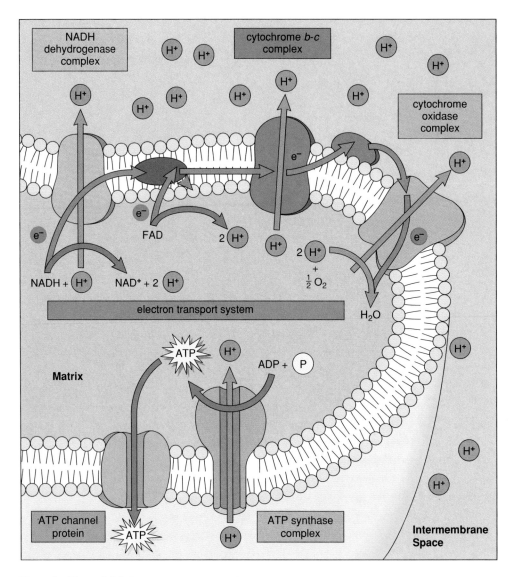

Organization of Cristae
Figure 7.8

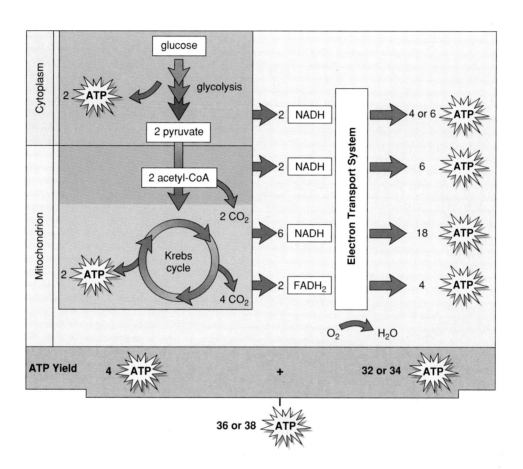

Energy Yield Per Glucose Breakdown
Figure 7.9

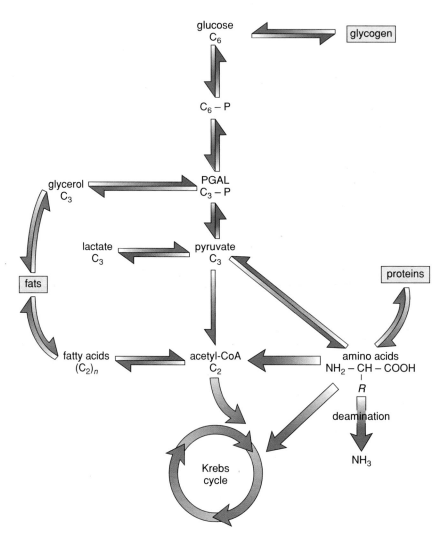

The Metabolic Pool Concept
Figure 7.10

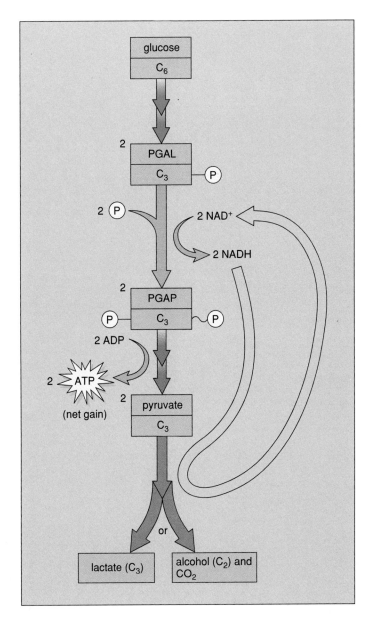

Fermentation
Figure 7.11

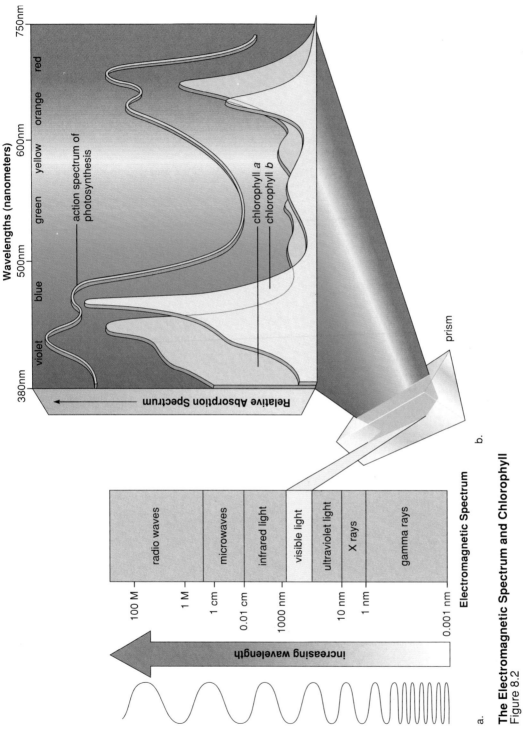

Wavelengths (nanometers)

Relative Absorption Spectrum

action spectrum of photosynthesis

chlorophyll *a*
chlorophyll *b*

violet   blue   green   yellow   orange   red

380nm   500nm   600nm   750nm

prism

b.

Electromagnetic Spectrum

radio waves

microwaves

infrared light

visible light

ultraviolet light

X rays

gamma rays

100 M

1 M

1 cm

0.01 cm

1000 nm

10 nm

1 nm

0.001 nm

increasing wavelength

a.

The Electromagnetic Spectrum and Chlorophyll
Figure 8.2

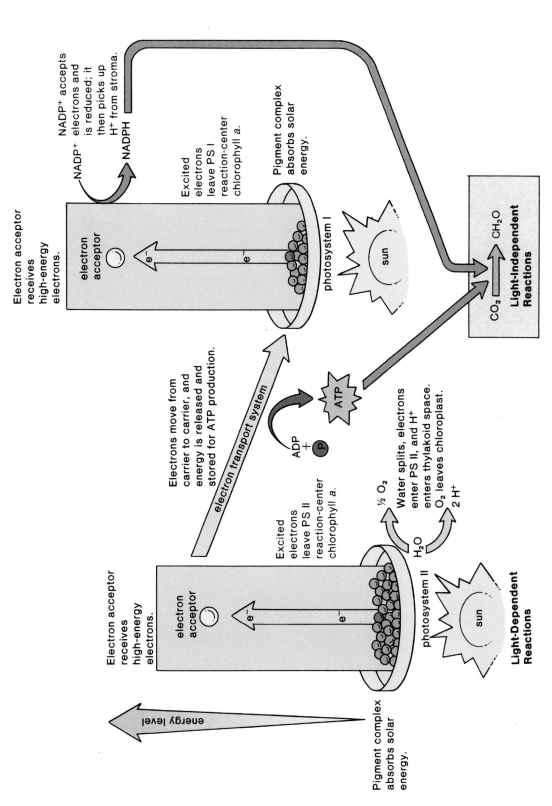

The Light-Dependent Reactions: The Noncyclic Electron Pathway
Figure 8.7

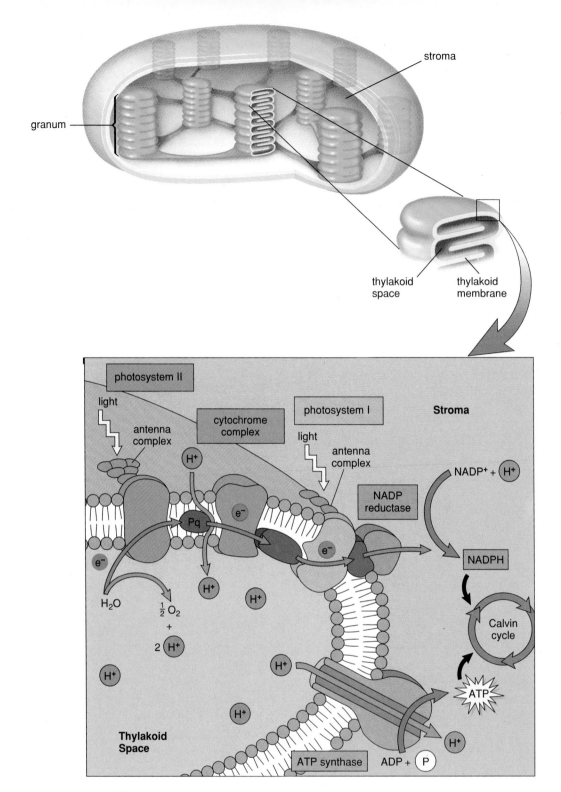

Organization of Thylakoid
Figure 8.8

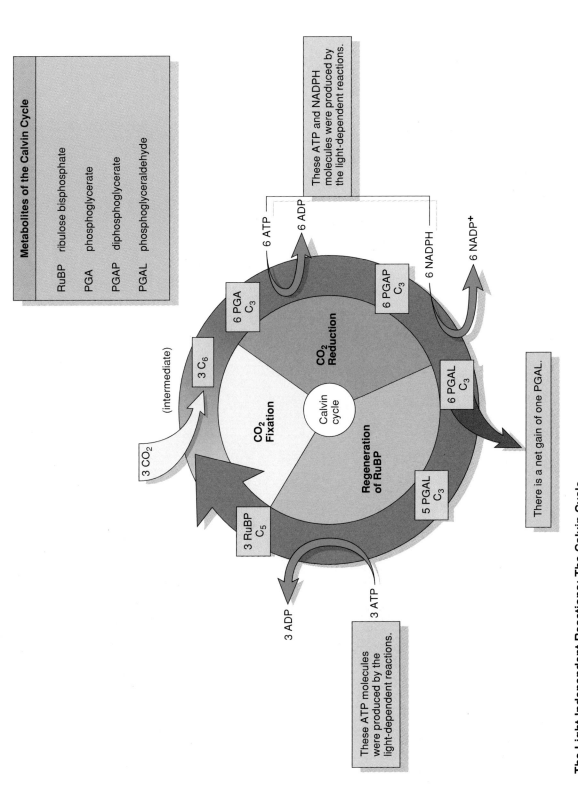

The Light-Independent Reactions: The Calvin Cycle
Figure 8.10

Metabolites of the Calvin Cycle

RuBP ribulose bisphosphate

PGA phosphoglycerate

PGAP diphosphoglycerate

PGAL phosphoglyceraldehyde

These ATP and NADPH molecules were produced by the light-dependent reactions.

6 ATP

6 ADP

6 NADPH

6 NADP⁺

6 PGA
C₃

3 C₆

(intermediate)

CO₂ Fixation

CO₂ Reduction

Calvin cycle

6 PGAP
C₃

3 CO₂

Regeneration of RuBP

6 PGAL
C₃

There is a net gain of one PGAL.

3 RuBP
C₅

5 PGAL
C₃

3 ADP

3 ATP

These ATP molecules were produced by the light-dependent reactions.

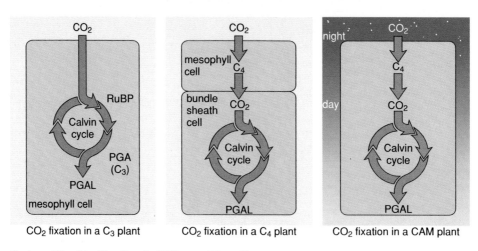

Carbon Dioxide Fixation in Different Plant Types
Figure 8.12

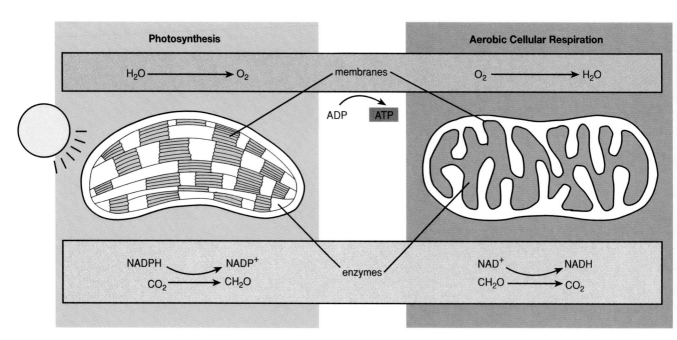

Photosynthesis vs. Aerobic Cellular Respiration
Figure 8.13

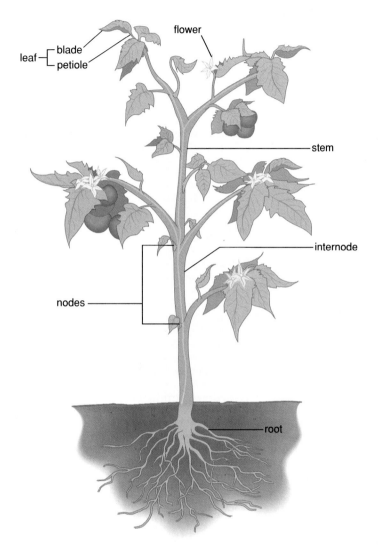

leaf ⌈ blade
     ⌊ petiole

flower

stem

internode

nodes

root

Organization of the Plant Body
Figure 9.1

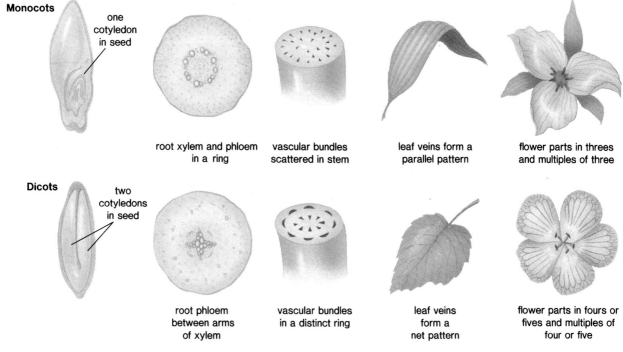

Monocots

one cotyledon in seed

root xylem and phloem in a ring

vascular bundles scattered in stem

leaf veins form a parallel pattern

flower parts in threes and multiples of three

Dicots

two cotyledons in seed

root phloem between arms of xylem

vascular bundles in a distinct ring

leaf veins form a net pattern

flower parts in fours or fives and multiples of four or five

Monocots vs. Dicots
Figure 9.2

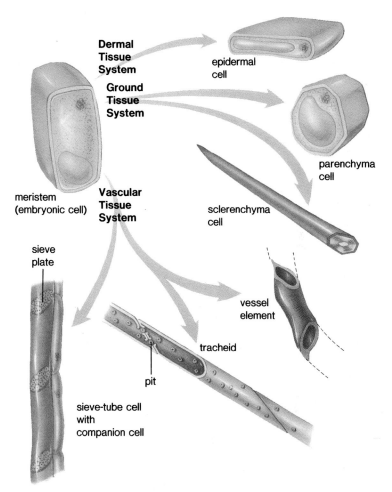

Dermal Tissue System

epidermal cell

Ground Tissue System

parenchyma cell

meristem (embryonic cell)

Vascular Tissue System

sclerenchyma cell

sieve plate

vessel element

tracheid

pit

sieve-tube cell with companion cell

Plant Cell Types
Figure 9.3

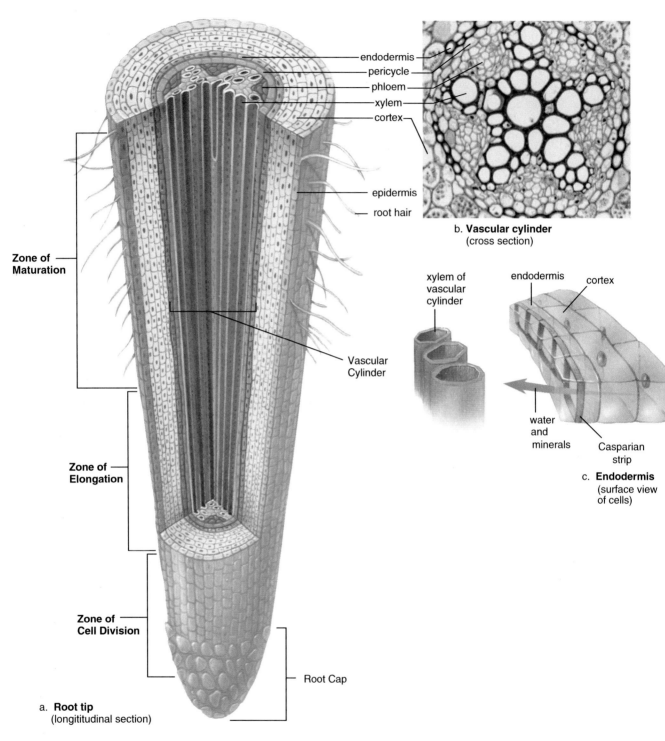

endodermis
pericycle
phloem
xylem
cortex

epidermis
root hair

b. Vascular cylinder
(cross section)

Zone of
Maturation

Zone of
Elongation

Zone of
Cell Division

Vascular
Cylinder

xylem of
vascular
cylinder

endodermis

cortex

water
and
minerals

Casparian
strip

c. **Endodermis**
(surface view
of cells)

Root Cap

a. **Root tip**
(longititudinal section)

Dicot Root Tip
Figure 9.7

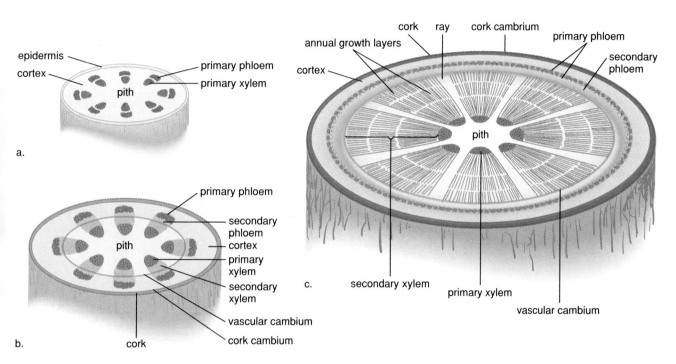

a.

epidermis
cortex
pith
primary phloem
primary xylem

b.

primary phloem
pith
secondary phloem
cortex
primary xylem
secondary xylem
vascular cambium
cork cambium
cork

c.

cork   ray   cork cambrium
annual growth layers
cortex
pith
primary phloem
secondary phloem
secondary xylem
primary xylem
vascular cambium

Secondary Growth in a Stem
Figure 9.14

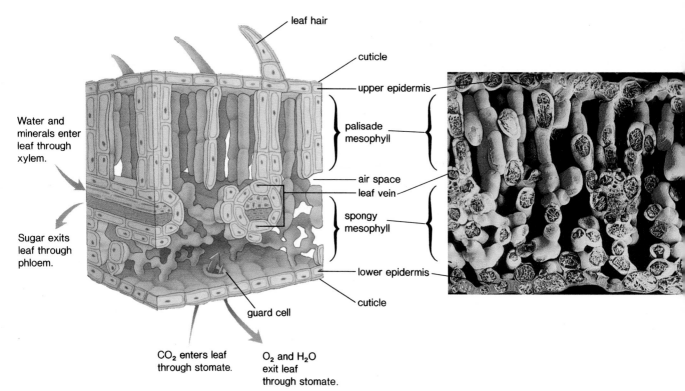

leaf hair

cuticle

upper epidermis

Water and
minerals enter
leaf through
xylem.

palisade
mesophyll

air space

leaf vein

spongy
mesophyll

Sugar exits
leaf through
phloem.

lower epidermis

cuticle

guard cell

CO_2 enters leaf
through stomate.

O_2 and H_2O
exit leaf
through stomate.

Leaf Structure
Figure 9.15

Stolon

Rhizome

root

rhizome
node
Tuber

papery
leaves
shoot
stem
Corm

a.

b.

c.

d.

Stem Modifications
Figure 9.16

57

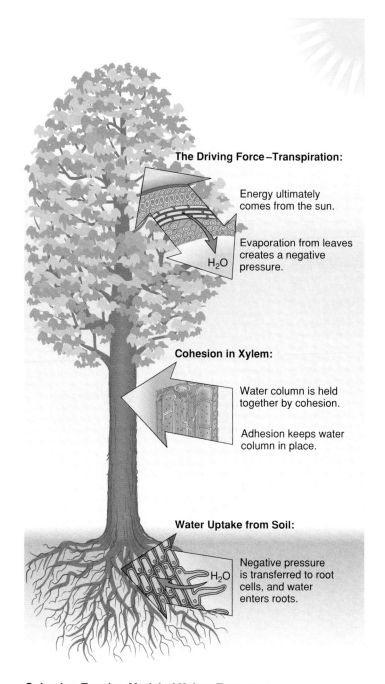

The Driving Force—Transpiration:

Energy ultimately comes from the sun.

Evaporation from leaves creates a negative pressure.

H_2O

Cohesion in Xylem:

Water column is held together by cohesion.

Adhesion keeps water column in place.

Water Uptake from Soil:

H_2O

Negative pressure is transferred to root cells, and water enters roots.

Cohesion-Tension Model of Xylem Transport
Figure 10.1

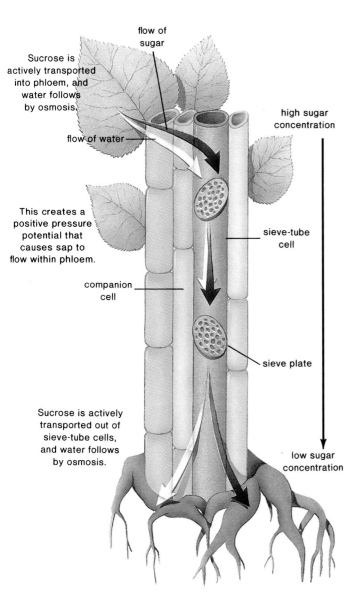

Sucrose is actively transported into phloem, and water follows by osmosis.

flow of sugar

flow of water

high sugar concentration

This creates a positive pressure potential that causes sap to flow within phloem.

sieve-tube cell

companion cell

sieve plate

Sucrose is actively transported out of sieve-tube cells, and water follows by osmosis.

low sugar concentration

Pressure-Flow Theory of Phloem Transport
Figure 10.3

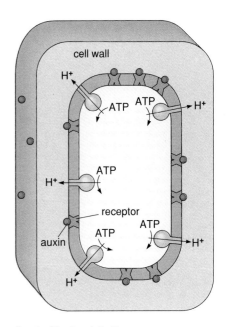

cell wall

H+

ATP ATP

H+

ATP

H+

receptor

ATP ATP

auxin

H+

H+

Auxin Mode of Action
Figure 10.6

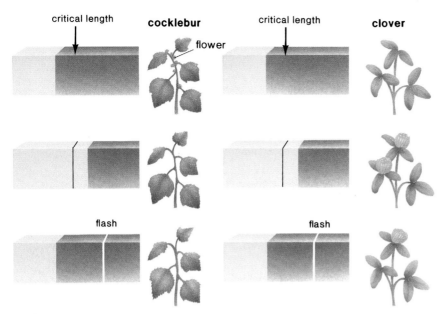

critical length   **cocklebur**   critical length   **clover**

flower

flash   flash

a. **Short-Day (Long-Night) Plant**   b. **Long-Day (Short-Night) Plant**

Day Length Effect on Two Types of Plants
Figure 10.7

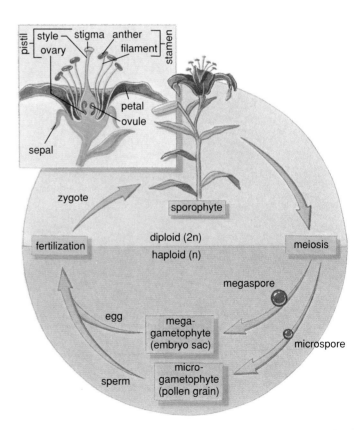

pistil — style   stigma   anther
         ovary   filament   stamen

                              petal
                              ovule

sepal

zygote   sporophyte

fertilization   diploid (2n)   meiosis
                 haploid (n)

                                          megaspore

egg   mega-
       gametophyte
       (embryo sac)   microspore

       micro-
       gametophyte
sperm (pollen grain)

Alternation of Generations in a Flowering Plant
Figure 10.8

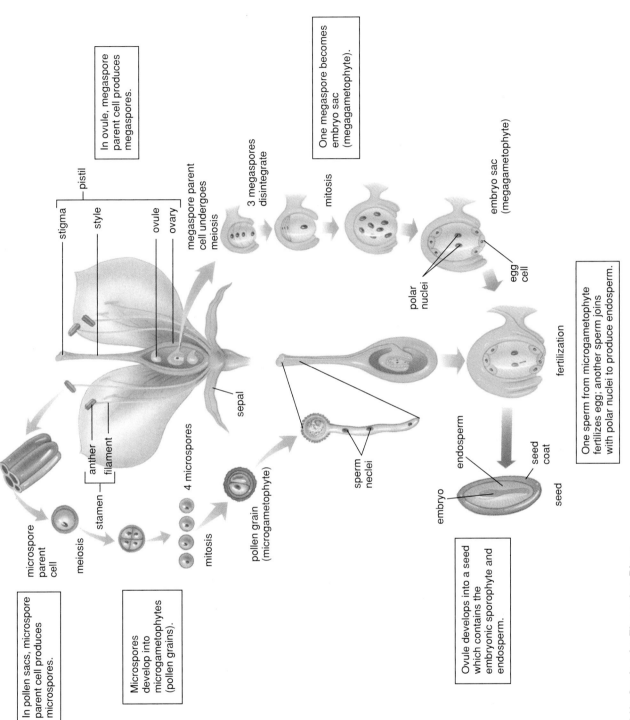

In pollen sacs, microspore parent cell produces microspores.

Microspores develop into microgametophytes (pollen grains).

In ovule, megaspore parent cell produces megaspores.

One megaspore becomes embryo sac (megagametophyte).

Ovule develops into a seed which contains the embryonic sporophyte and endosperm.

One sperm from microgametophyte fertilizes egg; another sperm joins with polar nuclei to produce endosperm.

stigma
style
ovule
ovary
pistil

megaspore parent cell undergoes meiosis

3 megaspores disintegrate

mitosis

embryo sac (megagametophyte)

polar nuclei

egg cell

sepal

anther
filament
stamen

microspore parent cell

meiosis

4 microspores

mitosis

pollen grain (microgametophyte)

sperm neclei

fertilization

endosperm

seed coat

embryo

seed

Life Cycle of a Flowering Plant
Figure 10.9

61

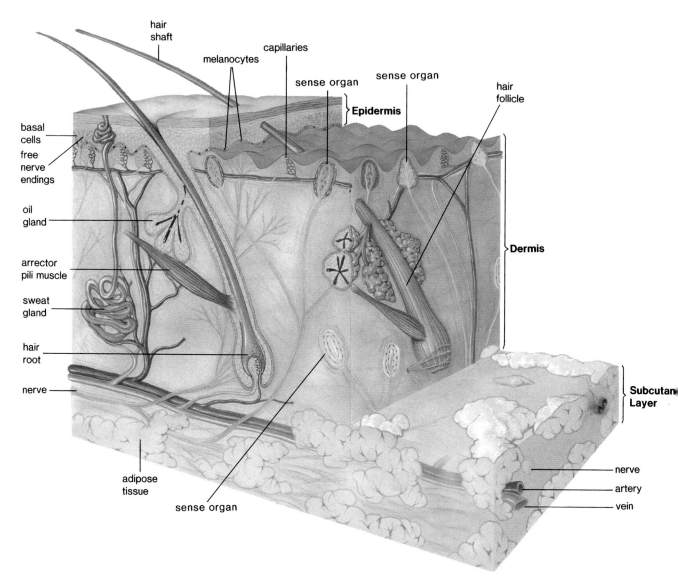

hair
shaft

melanocytes

capillaries

sense organ

sense organ

hair
follicle

} **Epidermis**

basal
cells

free
nerve
endings

oil
gland

arrector
pili muscle

sweat
gland

hair
root

nerve

adipose
tissue

sense organ

} **Dermis**

Subcutan
Layer

nerve

artery

vein

Human Skin Anatomy
Figure 11.7

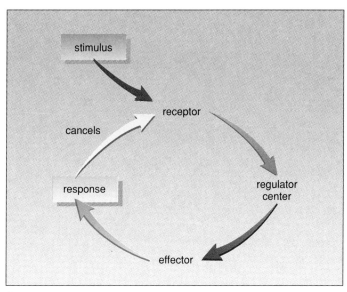

a.

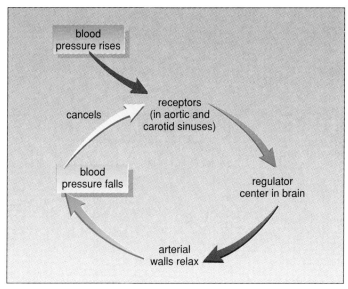

b.

Negative Feedback Control
Figure 11.10

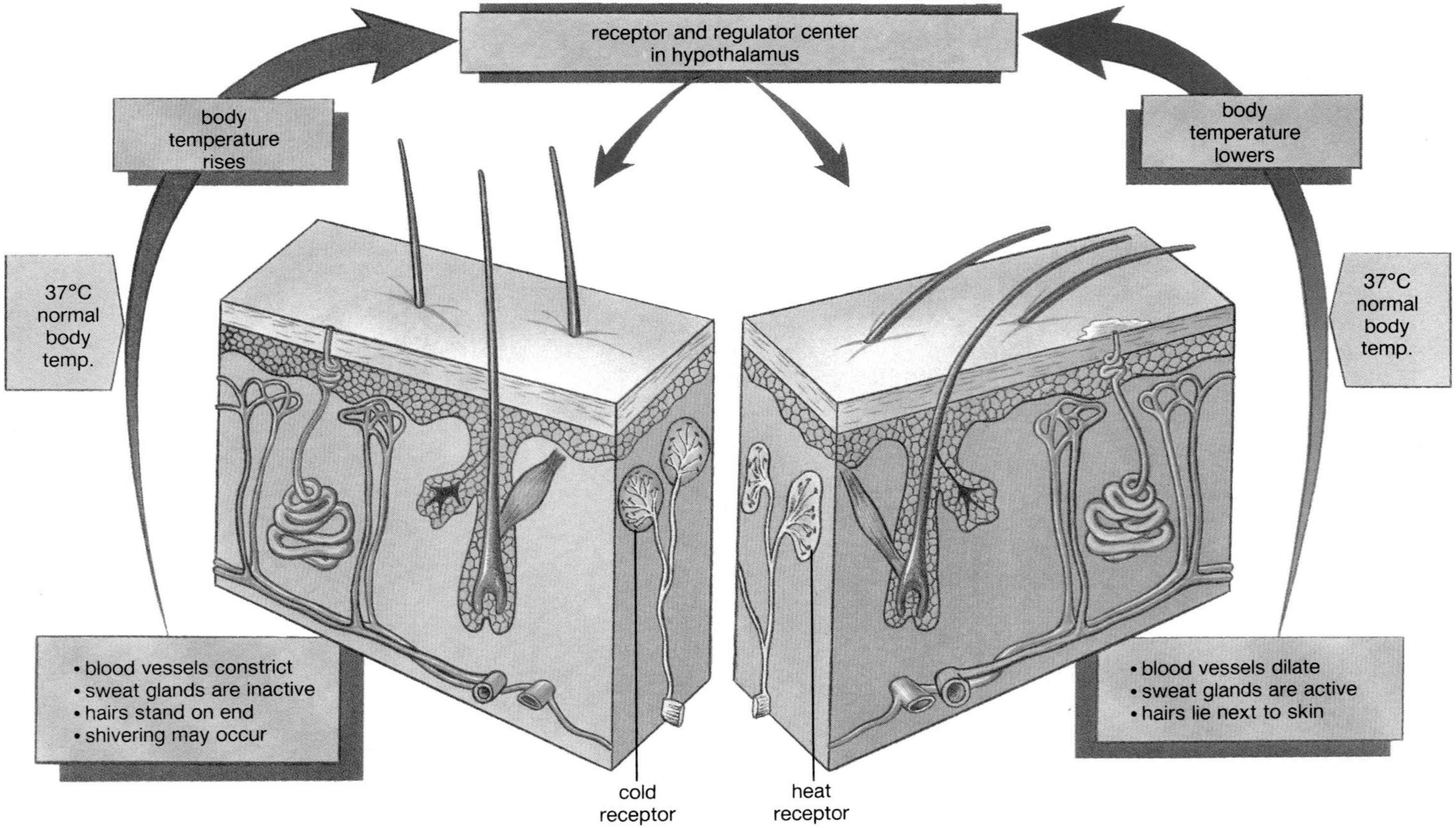

receptor and regulator center in hypothalamus
body temperature rises
37°C normal body temp.
• blood vessels constrict
• sweat glands are inactive
• hairs stand on end
• shivering may occur
cold receptor
heat receptor
body temperature lowers
37°C normal body temp.
• blood vessels dilate
• sweat glands are active
• hairs lie next to skin

Temperature Control
Figure 11.11

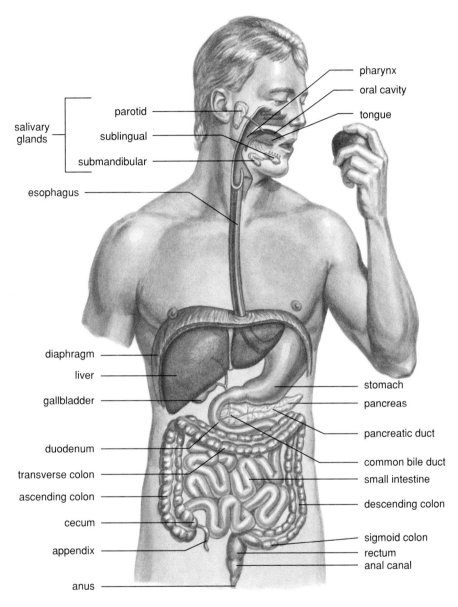

pharynx

oral cavity

tongue

salivary glands

parotid

sublingual

submandibular

esophagus

diaphragm

liver

gallbladder

duodenum

transverse colon

ascending colon

cecum

appendix

anus

stomach

pancreas

pancreatic duct

common bile duct

small intestine

descending colon

sigmoid colon

rectum

anal canal

Digestive System
Figure 12.1

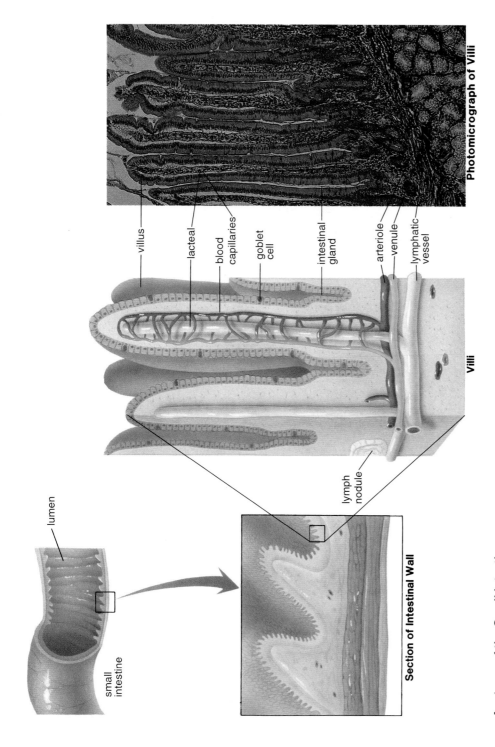

villus

lacteal

blood
capillaries

goblet
cell

intestinal
gland

arteriole

venule

lymphatic
vessel

Photomicrograph of Villi

Villi

lumen

small
intestine

lymph
nodule

Section of Intestinal Wall

Anatomy of the Small Intestine
Figure 12.6

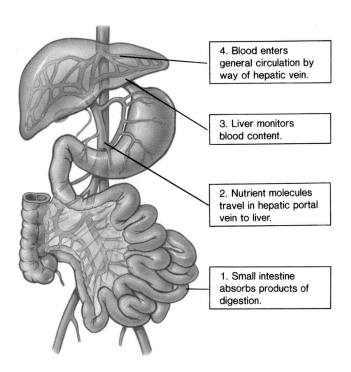

4. Blood enters general circulation by way of hepatic vein.

3. Liver monitors blood content.

2. Nutrient molecules travel in hepatic portal vein to liver.

1. Small intestine absorbs products of digestion.

Hepatic Portal System
Figure 12.11

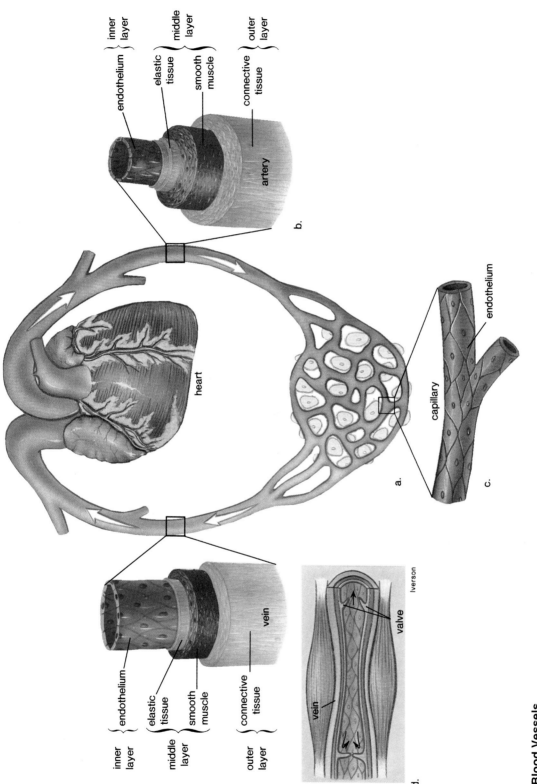

Blood Vessels
Figure 13.1

68

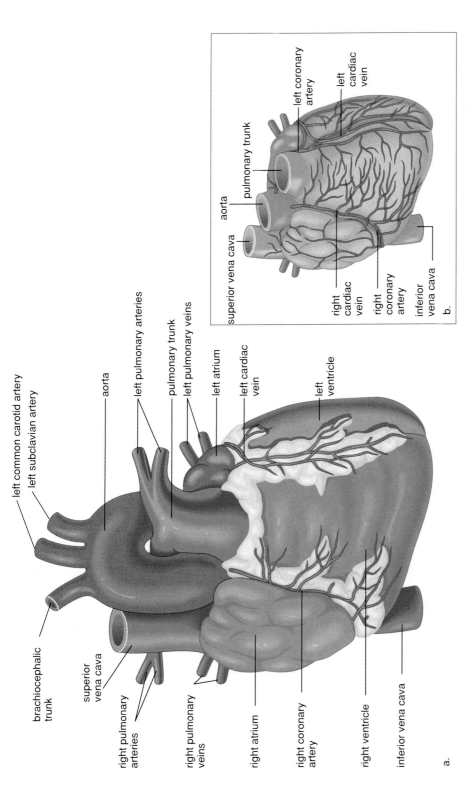

left common carotid artery

left subclavian artery

aorta

left pulmonary arteries

pulmonary trunk

left pulmonary veins

left atrium

left cardiac vein

left ventricle

brachiocephalic trunk

superior vena cava

right pulmonary arteries

right pulmonary veins

right atrium

right coronary artery

right ventricle

inferior vena cava

a.

left coronary artery

left cardiac vein

pulmonary trunk

aorta

superior vena cava

right cardiac vein

right coronary artery

inferior vena cava

b.

External Heart Anatomy
Figure 13.3

69

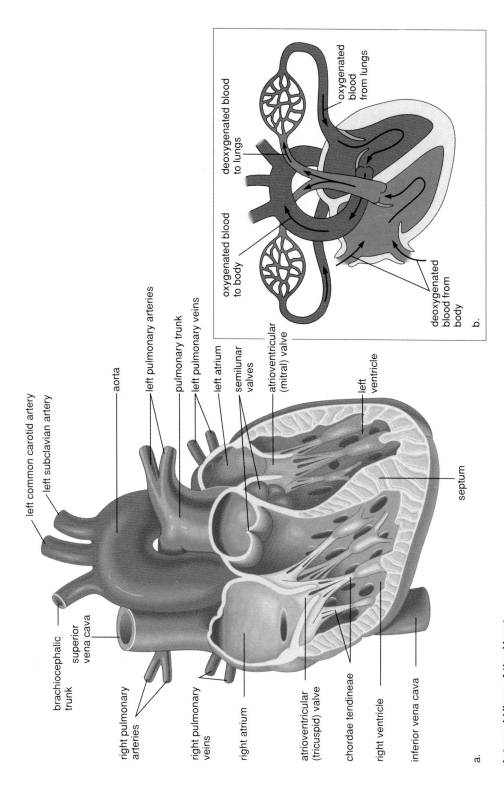

left common carotid artery
left subclavian artery

aorta

left pulmonary arteries
pulmonary trunk
left pulmonary veins
left atrium
semilunar valves
atrioventricular (mitral) valve

left ventricle

brachiocephalic trunk
superior vena cava

right pulmonary arteries

right pulmonary veins

right atrium

atrioventricular (tricuspid) valve

chordae tendineae

right ventricle

inferior vena cava

septum

a.

deoxygenated blood to lungs

oxygenated blood from lungs

oxygenated blood to body

deoxygenated blood from body

b.

Internal View of the Heart
Figure 13.4

70

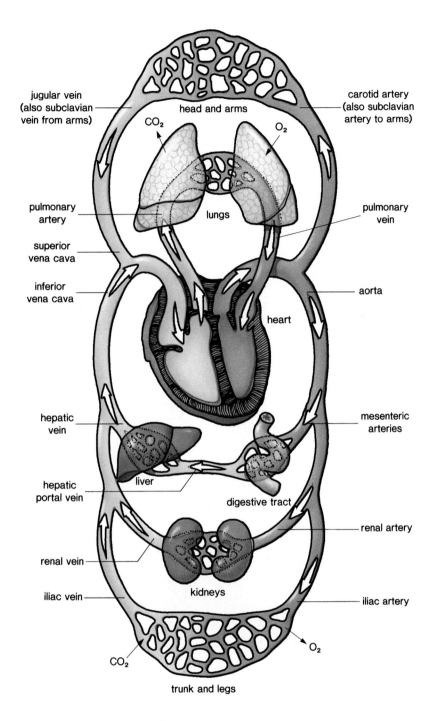

jugular vein
(also subclavian
vein from arms)

head and arms

carotid artery
(also subclavian
artery to arms)

CO_2

O_2

pulmonary
artery

lungs

pulmonary
vein

superior
vena cava

inferior
vena cava

heart

aorta

hepatic
vein

mesenteric
arteries

hepatic
portal vein

liver

digestive tract

renal vein

renal artery

iliac vein

kidneys

iliac artery

CO_2

O_2

trunk and legs

Cardiovascular System
Figure 13.8

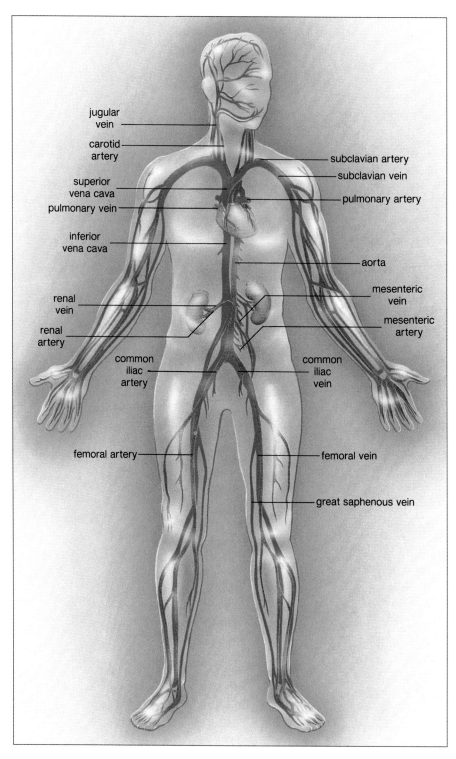

Human Circulatory System
Figure 13.9

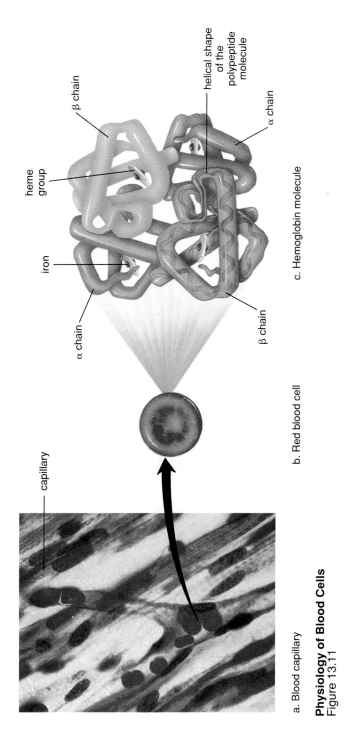

capillary

β chain

heme group

iron

α chain

helical shape of the polypeptide molecule

α chain

β chain

a. Blood capillary

b. Red blood cell

c. Hemoglobin molecule

Physiology of Blood Cells
Figure 13.11

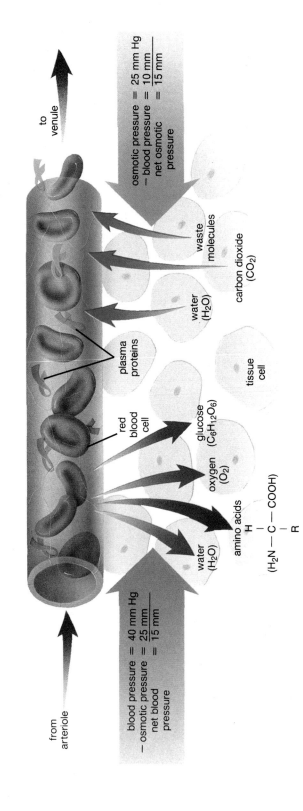

Exchange Between Blood and Tissue Across a Capillary Wall
Figure 13.15

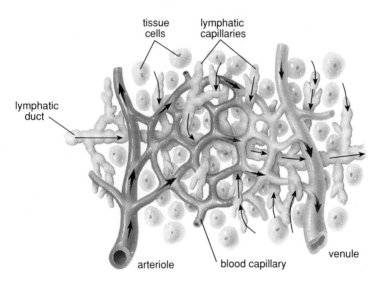

tissue
cells

lymphatic
capillaries

lymphatic
duct

arteriole

blood capillary

venule

Lymphatic Vessels
Figure 13.16

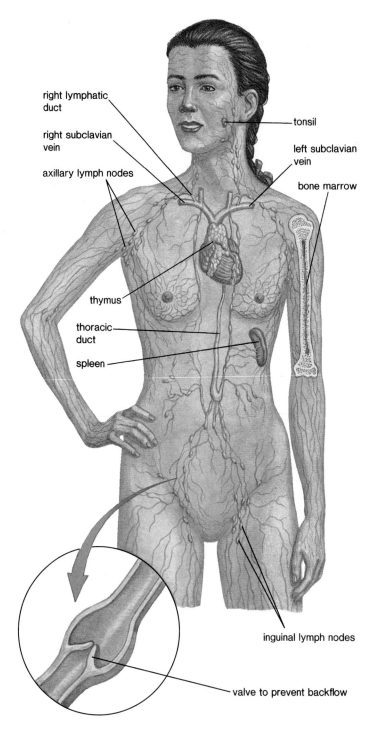

right lymphatic duct

tonsil

right subclavian vein

left subclavian vein

axillary lymph nodes

bone marrow

thymus

thoracic duct

spleen

inguinal lymph nodes

valve to prevent backflow

Lymphatic Vessel

Lymphatic System
Figure 14.1

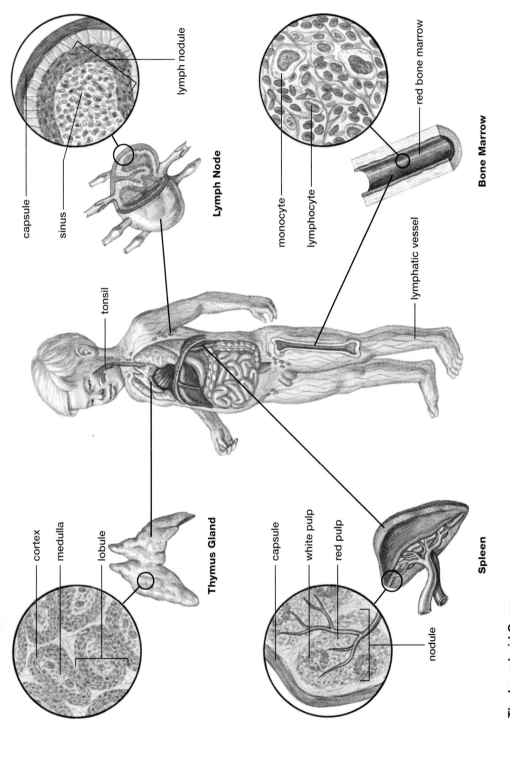

capsule

sinus

lymph nodule

Lymph Node

monocyte

lymphocyte

red bone marrow

Bone Marrow

tonsil

lymphatic vessel

cortex

medulla

lobule

Thymus Gland

capsule

white pulp

red pulp

nodule

Spleen

The Lymphoid Organs
Figure 14.2

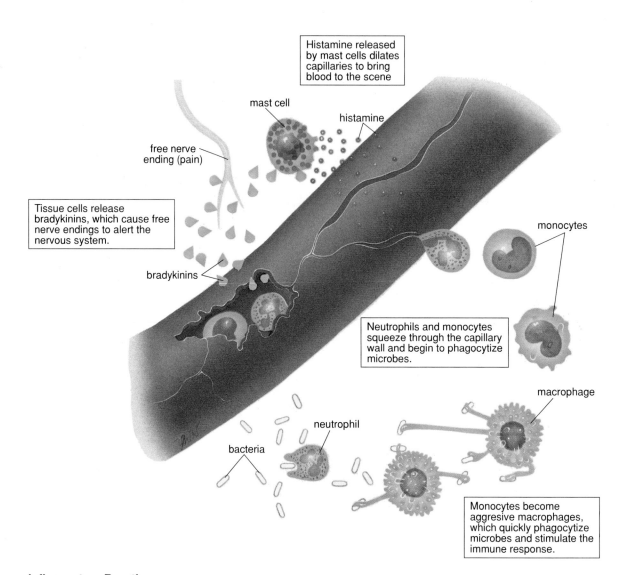

Histamine released by mast cells dilates capillaries to bring blood to the scene

mast cell

histamine

free nerve ending (pain)

Tissue cells release bradykinins, which cause free nerve endings to alert the nervous system.

bradykinins

monocytes

Neutrophils and monocytes squeeze through the capillary wall and begin to phagocytize microbes.

macrophage

bacteria

neutrophil

Monocytes become aggresive macrophages, which quickly phagocytize microbes and stimulate the immune response.

Inflammatory Reaction
Figure 14.4

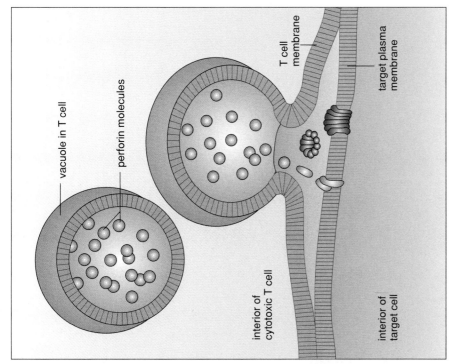

vacuole in T cell

perforin molecules

T cell membrane

target plasma membrane

interior of cytotoxic T cell

interior of target cell

b.

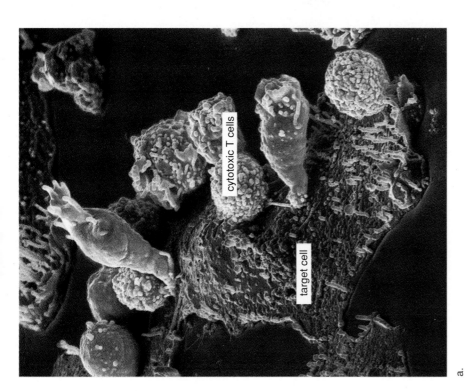

cytotoxic T cells

target cell

a.

Cell-Mediated Immunity
Figure 14.8

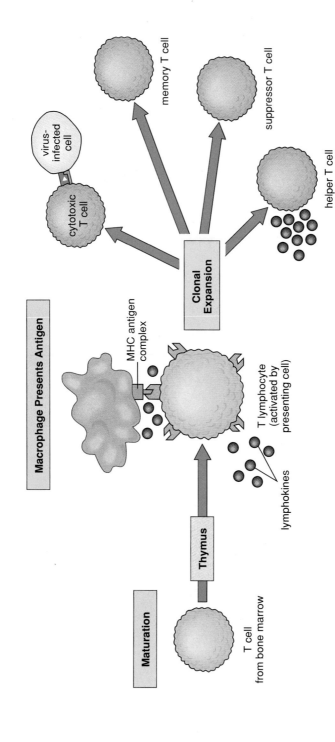

Activation and Diversity of T Cells
Figure 14.9

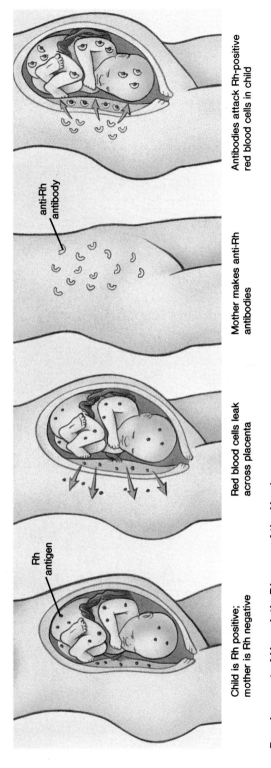

Child is Rh positive;
mother is Rh negative

Red blood cells leak
across placenta

Mother makes anti-Rh
antibodies

Antibodies attack Rh-positive
red blood cells in child

Rh
antigen

anti-Rh
antibody

Development of Hemolytic Disease of the Newborn
Figure 14.14

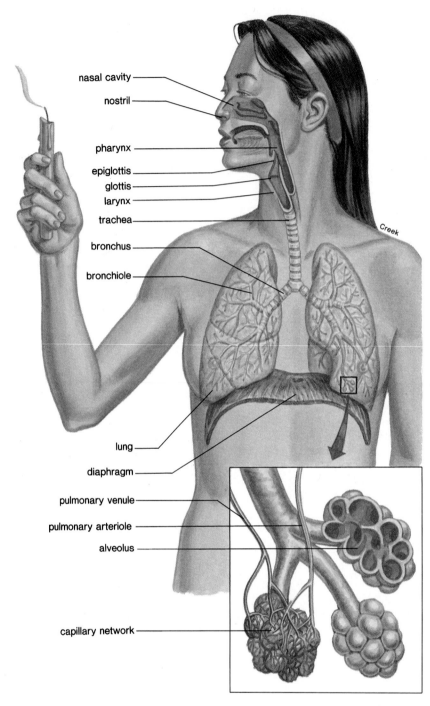

nasal cavity

nostril

pharynx

epiglottis

glottis

larynx

trachea

bronchus

bronchiole

lung

diaphragm

pulmonary venule

pulmonary arteriole

alveolus

capillary network

Creek

The Respiratory Tract
Figure 15.1

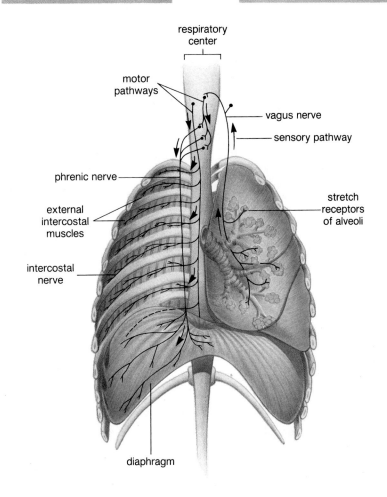

These structures are to be associated with inspiration.

These structures are to be associated with expiration.

respiratory center

motor pathways

vagus nerve

sensory pathway

phrenic nerve

external intercostal muscles

intercostal nerve

stretch receptors of alveoli

diaphragm

Nervous Control of Breathing
Figure 15.7

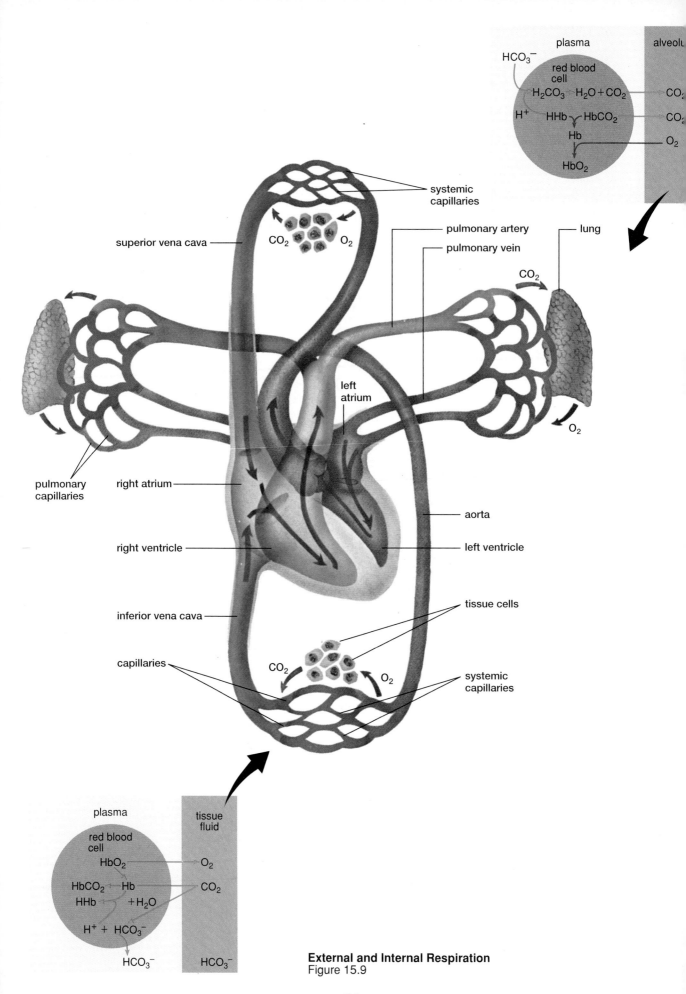

External and Internal Respiration
Figure 15.9

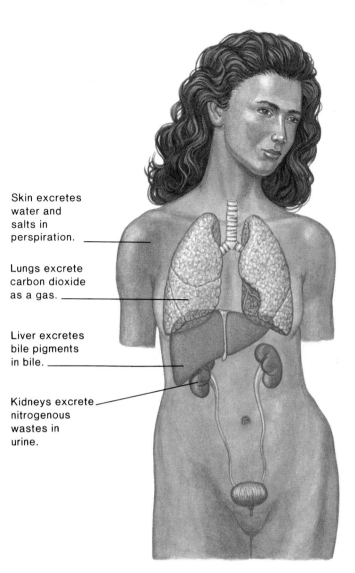

Skin excretes water and salts in perspiration.

Lungs excrete carbon dioxide as a gas.

Liver excretes bile pigments in bile.

Kidneys excrete nitrogenous wastes in urine.

Organs of Excretion
Figure 16.1

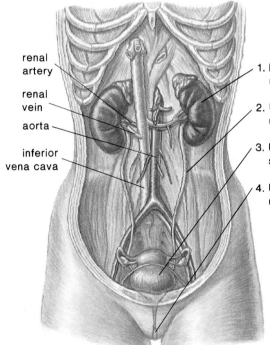

renal artery

renal vein

aorta

inferior vena cava

1. Kidneys produce urine.

2. Ureters transport urine.

3. Urinary bladder stores urine.

4. Urethra passes urine to outside.

The Urinary System
Figure 16.2

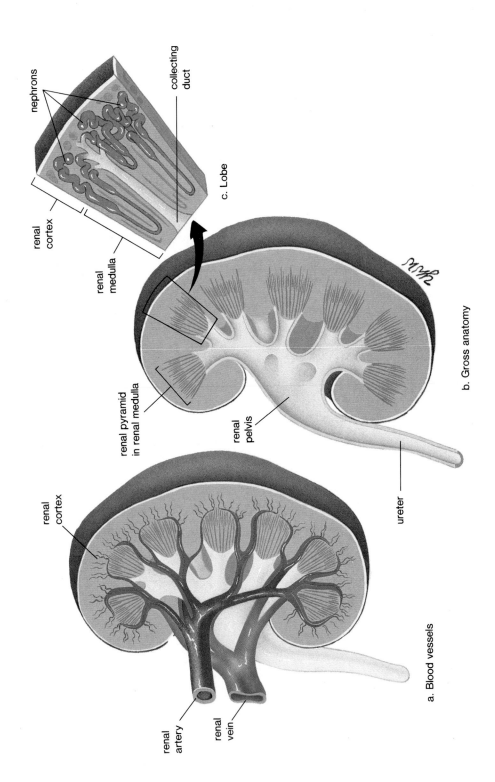

nephrons

collecting duct

c. Lobe

renal cortex

renal medulla

renal pyramid in renal medulla

renal cortex

renal pelvis

ureter

renal artery

renal vein

b. Gross anatomy

a. Blood vessels

Gross Anatomy of the Kidney
Figure 16.3

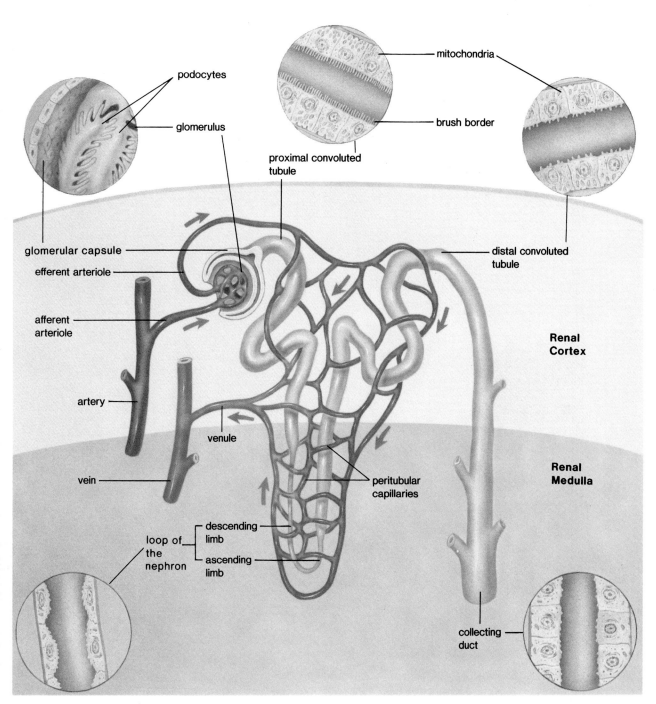

Nephron Microscopic and Macroscopic Anatomy
Figure 16.4

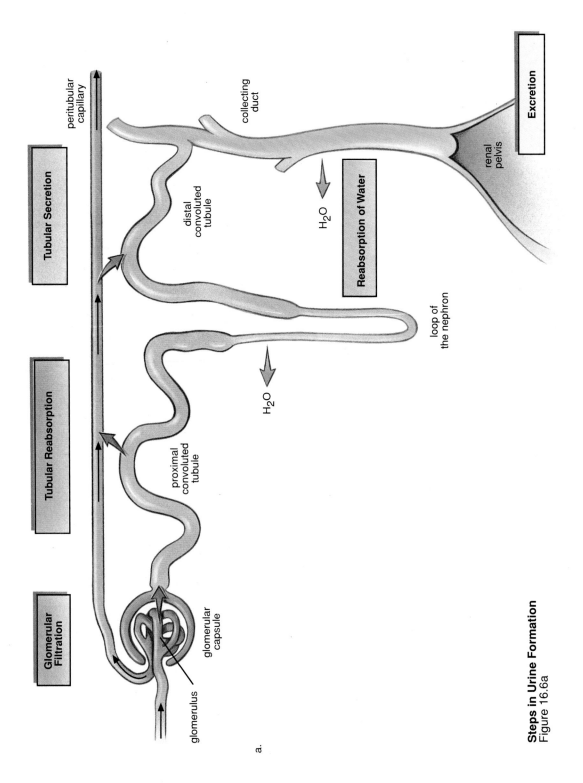

Glomerular Filtration

Tubular Reabsorption

Tubular Secretion

peritubular capillary

glomerulus

glomerular capsule

proximal convoluted tubule

H₂O

distal convoluted tubule

collecting duct

H₂O

Reabsorption of Water

loop of the nephron

renal pelvis

Excretion

a.

Steps in Urine Formation
Figure 16.6a

Steps in Urine Formation

Name	Process	Examples of Molecules
Glomerular filtration	Blood pressure forces small molecules from the glomerulus into the glomerular capsule.	Water, glucose, amino acids, salts, urea, uric acid, creatinine
Tubular reabsorption	Diffusion and active transport return molecules to blood at the proximal convoluted tubule.	Water, glucose, amino acids, salts
Tubular secretion	Active transport moves molecules from blood into the distal convoluted tubule.	Uric acid, creatinine, hydrogen ions, ammonia, penicillin
Reabsorption of water	Along the length of the nephron and notably at loop of the nephron and collecting duct, water returns by osmosis following active reabsorption of salt.	Salt, water
Excretion	Urine formation rids body of metabolic wastes.	Water, salts, urea, uric acid, ammonium, creatinine

b.

Steps in Urine Formation
Figure 16.6b

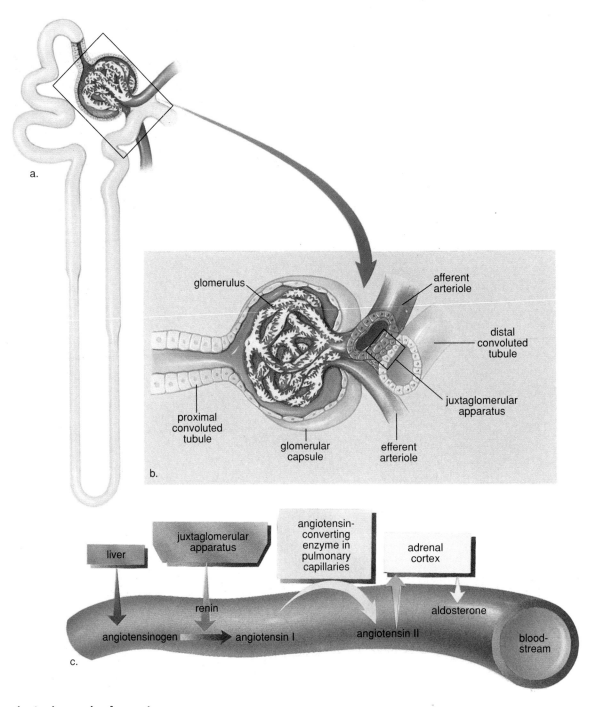

a.

glomerulus

afferent
arteriole

distal
convoluted
tubule

juxtaglomerular
apparatus

proximal
convoluted
tubule

glomerular
capsule

efferent
arteriole

b.

liver

juxtaglomerular
apparatus

angiotensin-
converting
enzyme in
pulmonary
capillaries

adrenal
cortex

renin

aldosterone

angiotensinogen

angiotensin I

angiotensin II

blood-
stream

c.

Juxtaglomerular Apparatus
Figure 16.8

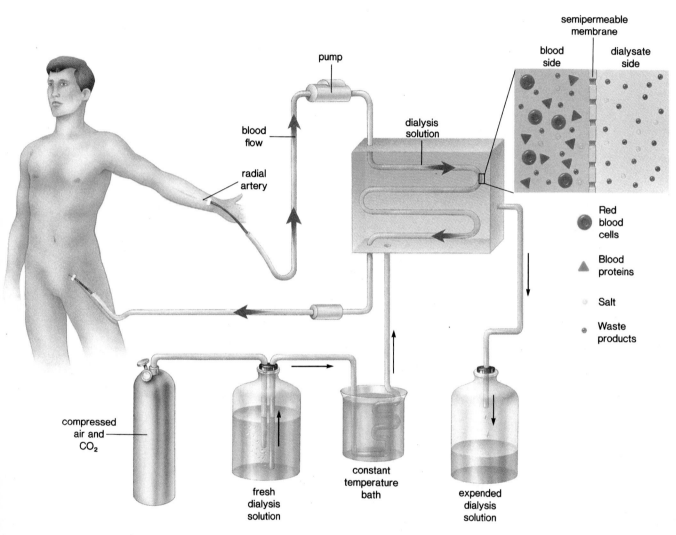

semipermeable
membrane

blood side

dialysate side

pump

blood flow

radial artery

dialysis solution

Red blood cells

Blood proteins

Salt

Waste products

compressed air and CO_2

fresh dialysis solution

constant temperature bath

expended dialysis solution

An Artificial Kidney Machine
Figure 16.10

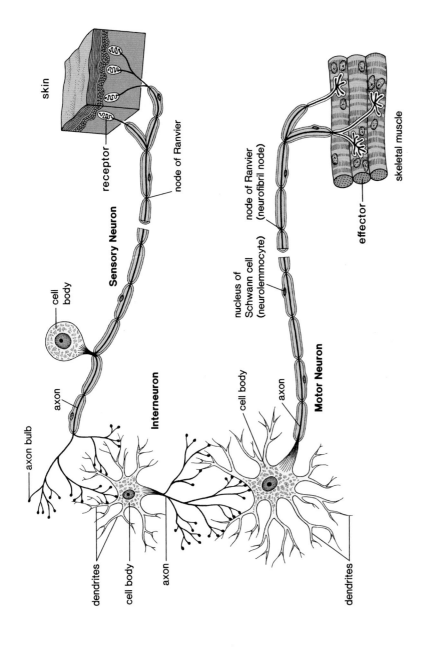

Types of Neurons
Figure 17.1

92

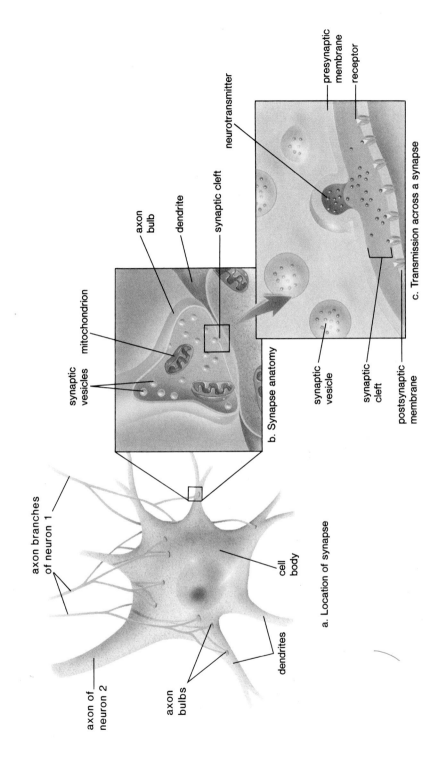

axon branches
of neuron 1

axon of
neuron 2

axon
bulbs

dendrites

cell
body

a. Location of synapse

synaptic
vesicles

mitochondrion

axon
bulb

dendrite

synaptic cleft

b. Synapse anatomy

neurotransmitter

presynaptic
membrane

receptor

synaptic
vesicle

synaptic
cleft

postsynaptic
membrane

c. Transmission across a synapse

Synapse Structure and Function
Figure 17.3

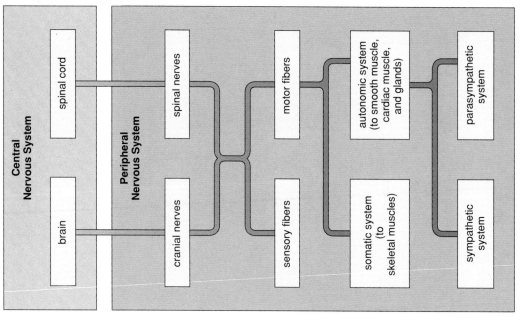

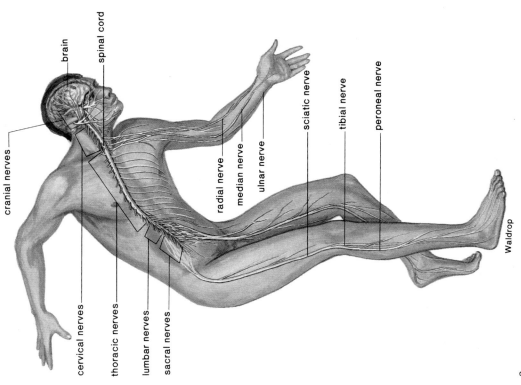

a.

b.

Peripheral Nervous System Compared With Central Nervous System
Figure 17.4

94

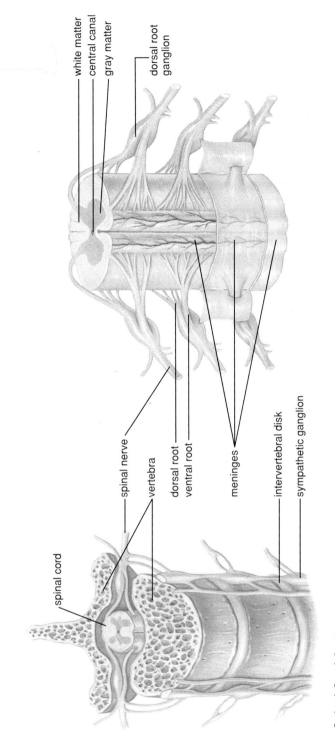

white matter
central canal
gray matter

dorsal root ganglion

spinal nerve

vertebra

dorsal root
ventral root

meninges

intervertebral disk

sympathetic ganglion

spinal cord

Spinal Cord Anatomy
Figure 17.5

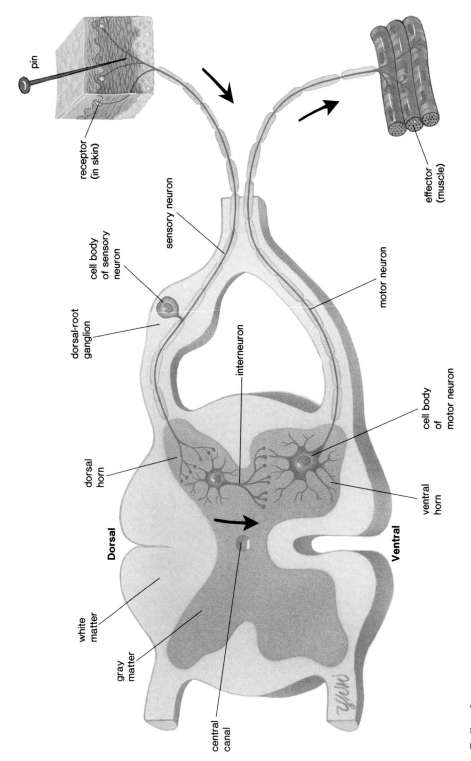

pin

receptor
(in skin)

sensory neuron

cell body
of sensory
neuron

dorsal-root
ganglion

interneuron

effector
(muscle)

motor neuron

cell body
of
motor neuron

ventral
horn

dorsal
horn

Dorsal

Ventral

white
matter

gray
matter

central
canal

Reflex Arc
Figure 17.6

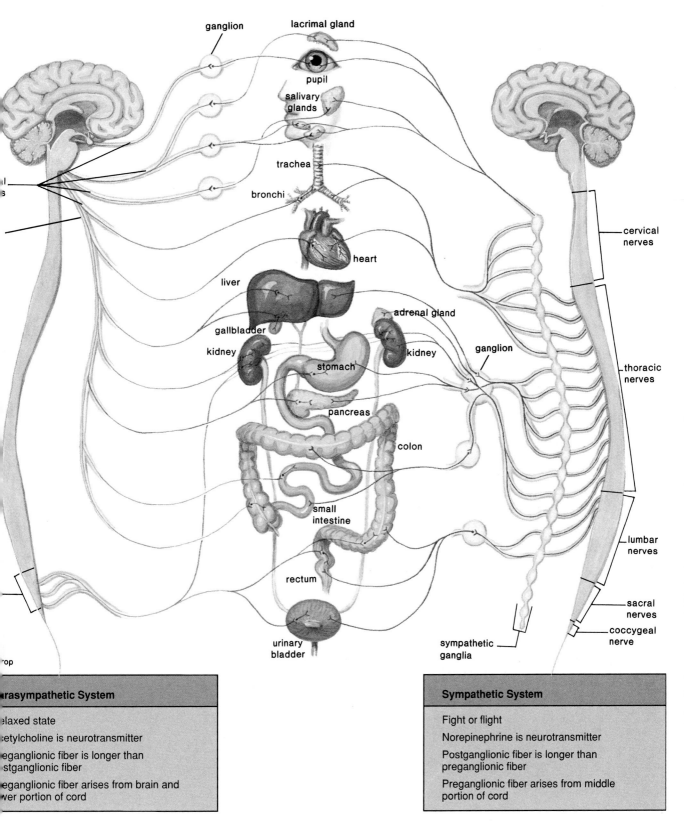

ganglion lacrimal gland

pupil

salivary
glands

trachea

bronchi

heart

liver

adrenal gland

gallbladder

kidney   kidney   ganglion

stomach

pancreas

colon

small
intestine

rectum

urinary
bladder

cervical
nerves

thoracic
nerves

lumbar
nerves

sacral
nerves

coccygeal
nerve

sympathetic
ganglia

op

rasympathetic System

elaxed state

cetylcholine is neurotransmitter

eganglionic fiber is longer than
stganglionic fiber

eganglionic fiber arises from brain and
ver portion of cord

Sympathetic System

Fight or flight

Norepinephrine is neurotransmitter

Postganglionic fiber is longer than
preganglionic fiber

Preganglionic fiber arises from middle
portion of cord

Autonomic Nervous System Structure and Function
Figure 17.7

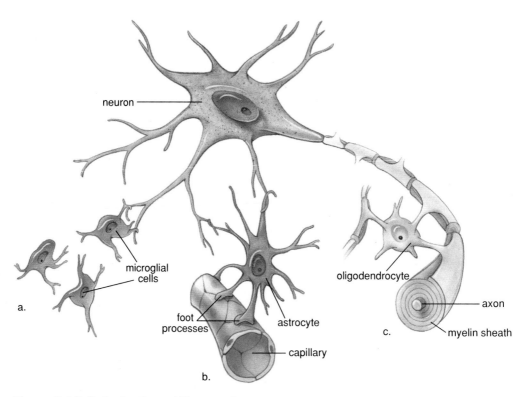

neuron

microglial
cells

a.

foot
processes

astrocyte

capillary

b.

oligodendrocyte

axon

c.

myelin sheath

Neuroglial Cells in the Central Nervous System
Figure 17.8

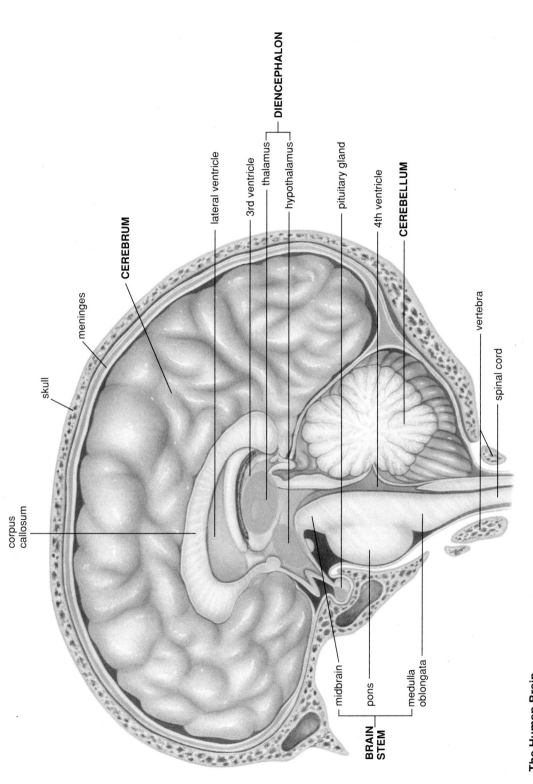

The Human Brain
Figure 17.9

corpus callosum

skull

meninges

CEREBRUM

lateral ventricle

3rd ventricle

thalamus

hypothalamus

DIENCEPHALON

pituitary gland

4th ventricle

CEREBELLUM

vertebra

spinal cord

midbrain

pons

medulla oblongata

BRAIN STEM

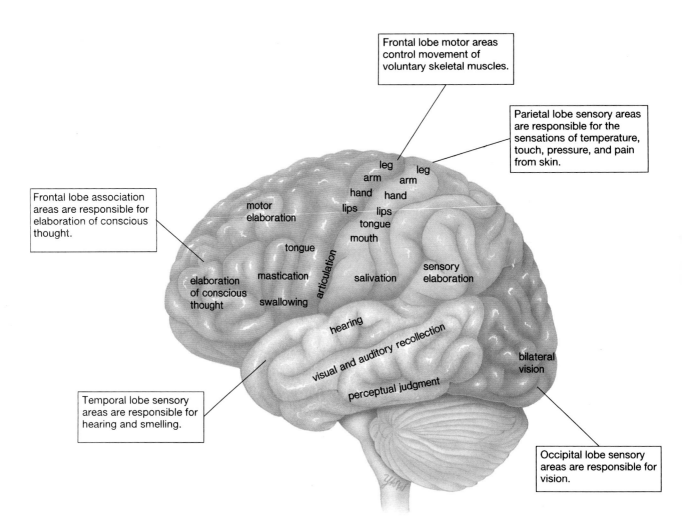

Frontal lobe motor areas control movement of voluntary skeletal muscles.

Parietal lobe sensory areas are responsible for the sensations of temperature, touch, pressure, and pain from skin.

Frontal lobe association areas are responsible for elaboration of conscious thought.

leg
arm
hand
lips
tongue
mouth

leg
arm
hand
lips

motor
elaboration

tongue

elaboration
of conscious
thought

mastication

swallowing

articulation

salivation

sensory
elaboration

hearing

visual and auditory recollection

perceptual judgment

bilateral
vision

Temporal lobe sensory areas are responsible for hearing and smelling.

Occipital lobe sensory areas are responsible for vision.

The Cerebral Cortex
Figure 17.10

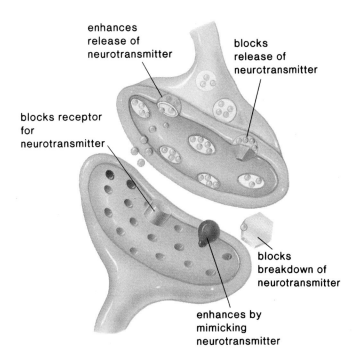

enhances
release of
neurotransmitter

blocks
release of
neurotransmitter

blocks receptor
for
neurotransmitter

blocks
breakdown of
neurotransmitter

enhances by
mimicking
neurotransmitter

Drug Action At a Synapse
Figure 17.12

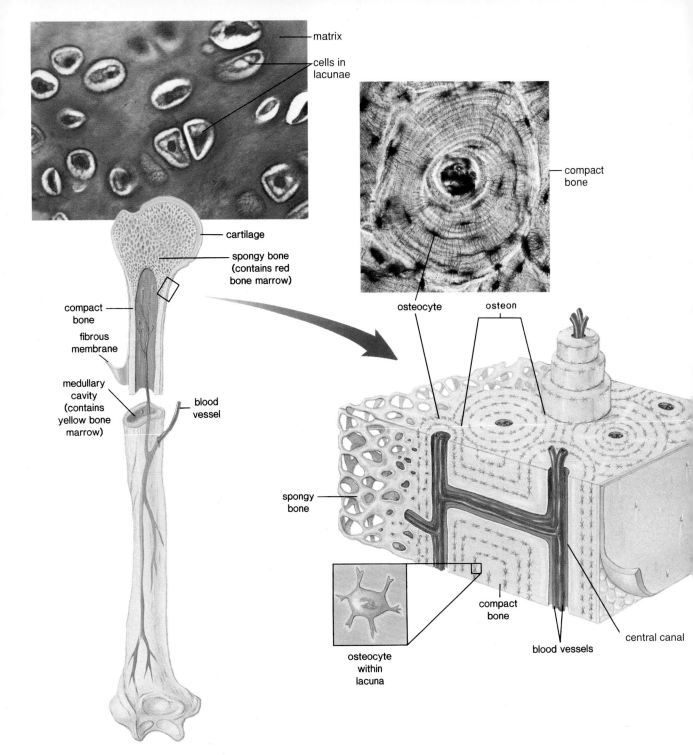

matrix

cells in
lacunae

compact
bone

cartilage

spongy bone
(contains red
bone marrow)

osteocyte   osteon

compact
bone

fibrous
membrane

medullary
cavity
(contains
yellow bone
marrow)

blood
vessel

spongy
bone

compact
bone

osteocyte
within
lacuna

blood vessels

central canal

Anatomy of a Long Bone
Figure 18.1

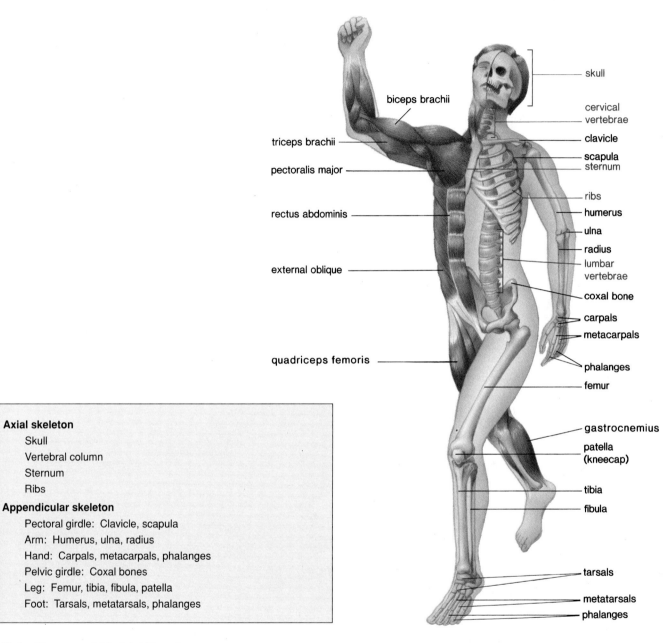

skull

cervical vertebrae

clavicle

scapula

sternum

ribs

humerus

ulna

radius

lumbar vertebrae

coxal bone

carpals

metacarpals

phalanges

femur

gastrocnemius

patella (kneecap)

tibia

fibula

tarsals

metatarsals

phalanges

biceps brachii

triceps brachii

pectoralis major

rectus abdominis

external oblique

quadriceps femoris

Axial skeleton

Skull

Vertebral column

Sternum

Ribs

Appendicular skeleton

Pectoral girdle: Clavicle, scapula

Arm: Humerus, ulna, radius

Hand: Carpals, metacarpals, phalanges

Pelvic girdle: Coxal bones

Leg: Femur, tibia, fibula, patella

Foot: Tarsals, metatarsals, phalanges

Major Bones and Skeletal Muscles
Figure 18.2

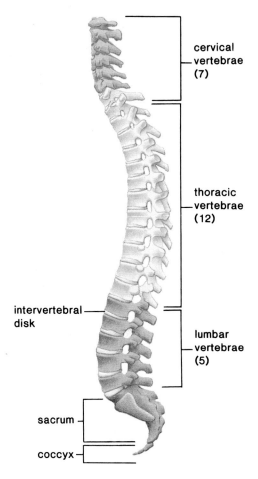

cervical
vertebrae
(7)

thoracic
vertebrae
(12)

intervertebral
disk

lumbar
vertebrae
(5)

sacrum

coccyx

The Vertebral Column
Figure 18.3

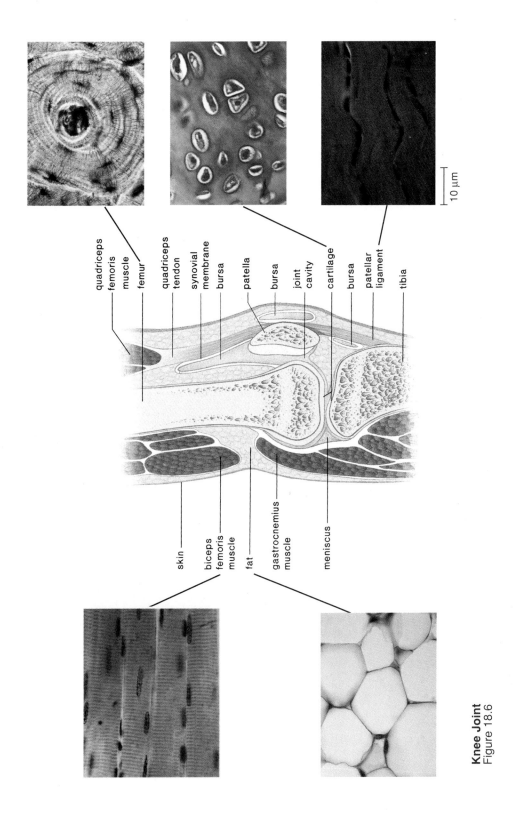

Knee Joint
Figure 18.6

10 μm

quadriceps femoris muscle
femur
quadriceps tendon
synovial membrane
bursa
patella
bursa
joint cavity
cartilage
bursa
patellar ligament
tibia

skin
biceps femoris muscle
fat
gastrocnemius muscle
meniscus

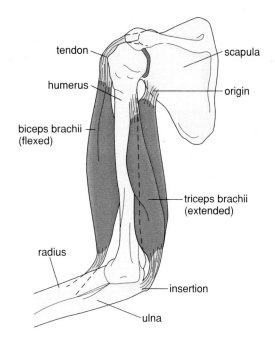

Attachment of Skeletal Muscles
Figure 18.7

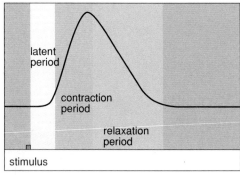

b.

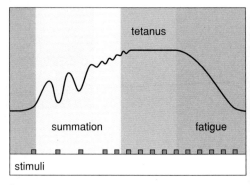

c.

Physiology of Skeletal Muscle Contraction
Figure 18.8

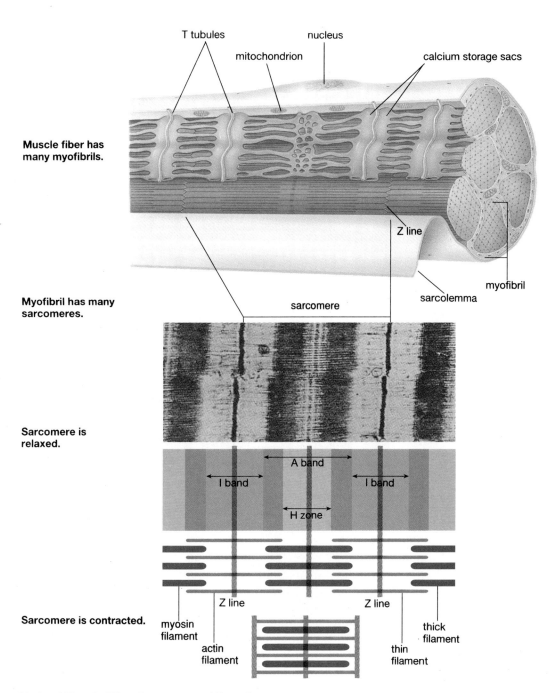

T tubules nucleus

mitochondrion calcium storage sacs

**Muscle fiber has
many myofibrils.**

Z line

myofibril

sarcolemma

**Myofibril has many
sarcomeres.**

sarcomere

**Sarcomere is
relaxed.**

A band

I band I band

H zone

Z line Z line

Sarcomere is contracted. myosin
filament thick
filament

actin
filament thin
filament

Skeletal Muscle Fiber Structure and Function
Figure 18.9

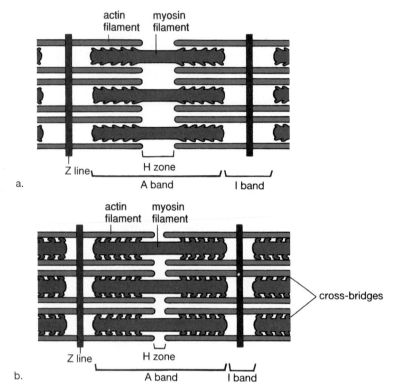

a.

b.

Sliding Filament Theory
Figure 18.10

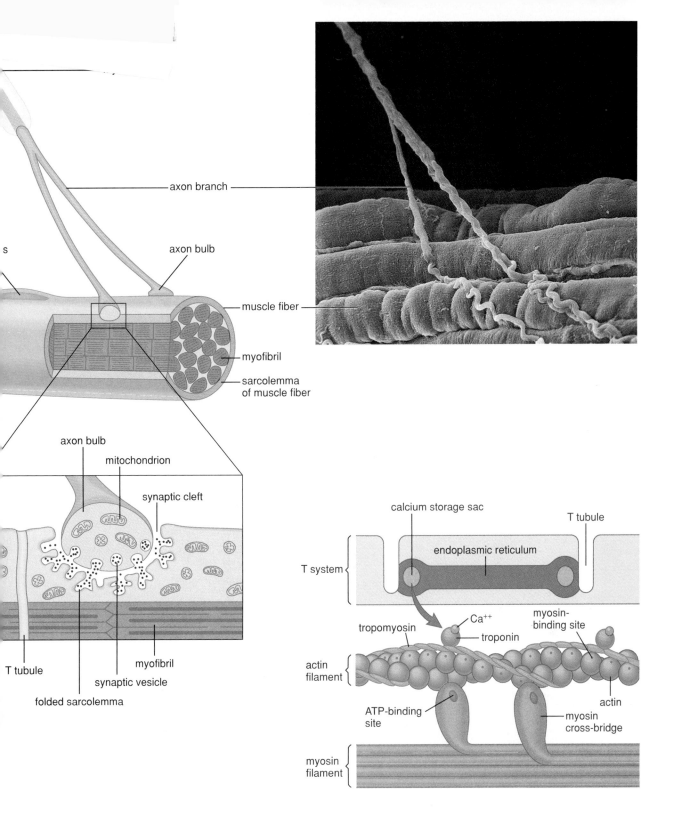

Neuromuscular Junction
Figure 18.11

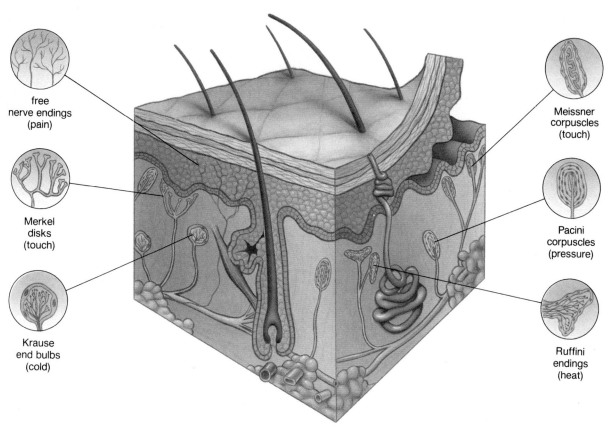

free
nerve endings
(pain)

Merkel
disks
(touch)

Krause
end bulbs
(cold)

Meissner
corpuscles
(touch)

Pacini
corpuscles
(pressure)

Ruffini
endings
(heat)

Receptors in Human Skin
Figure 19.2

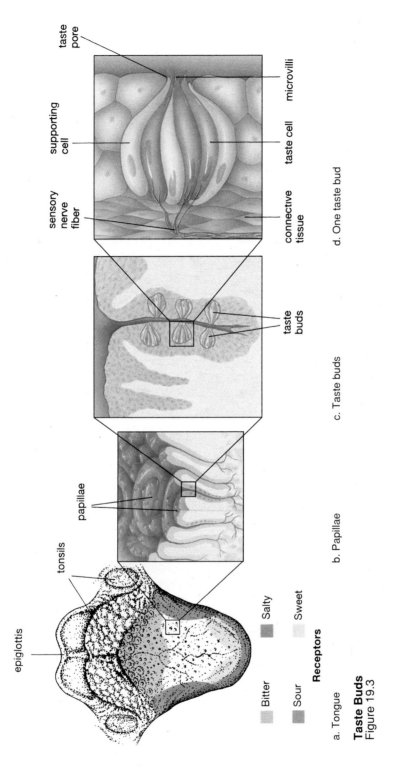

taste
pore

microvilli

supporting
cell

taste cell

sensory
nerve
fiber

connective
tissue

d. One taste bud

taste
buds

c. Taste buds

papillae

b. Papillae

tonsils

epiglottis

Receptors

Bitter

Salty

Sour

Sweet

a. Tongue

Taste Buds
Figure 19.3

111

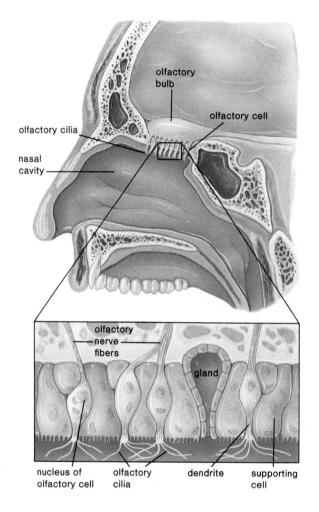

olfactory bulb

olfactory cell

olfactory cilia

nasal cavity

olfactory nerve fibers

gland

nucleus of olfactory cell

olfactory cilia

dendrite

supporting cell

Olfactory Cell Location and Anatomy
Figure 19.4

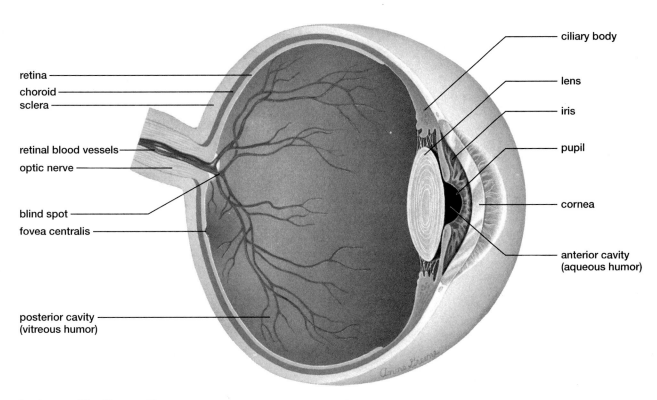

retina

choroid

sclera

retinal blood vessels

optic nerve

blind spot

fovea centralis

posterior cavity
(vitreous humor)

ciliary body

lens

iris

pupil

cornea

anterior cavity
(aqueous humor)

Anatomy of the Human Eye
Figure 19.6

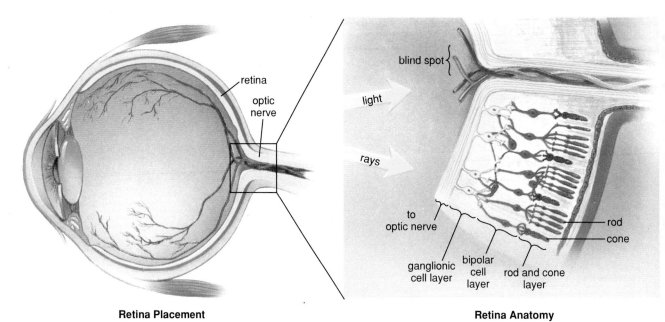

Retina Placement

Retina Anatomy

Anatomy of the Retina
Figure 19.7

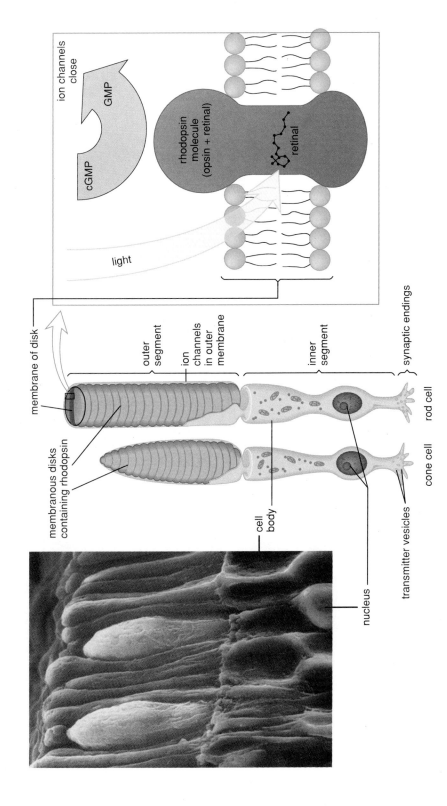

Structure and Function of Rods and Cones
Figure 19.10

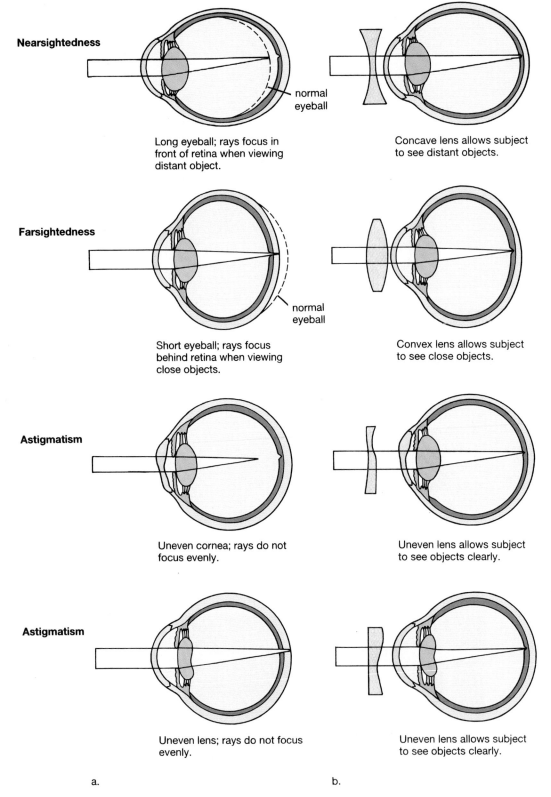

Nearsightedness

Long eyeball; rays focus in front of retina when viewing distant object.

normal eyeball

Concave lens allows subject to see distant objects.

Farsightedness

Short eyeball; rays focus behind retina when viewing close objects.

normal eyeball

Convex lens allows subject to see close objects.

Astigmatism

Uneven cornea; rays do not focus evenly.

Uneven lens allows subject to see objects clearly.

Astigmatism

Uneven lens; rays do not focus evenly.

Uneven lens allows subject to see objects clearly.

a.

b.

Common Abnormalities of the Eye
Figure 19.12

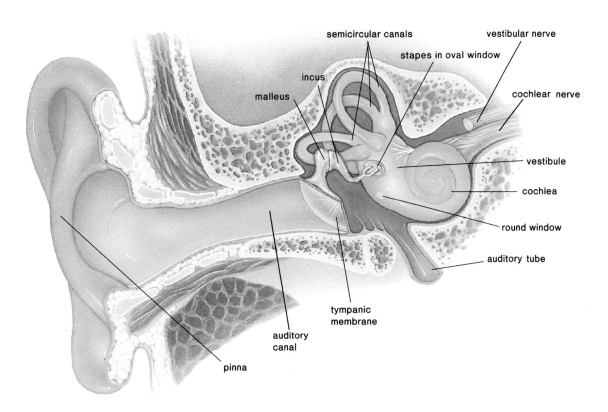

Anatomy of the Human Ear
Figure 19.13

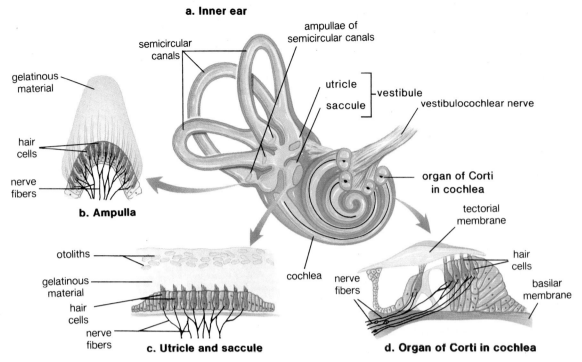

a. Inner ear

gelatinous
material

hair
cells

nerve
fibers

b. Ampulla

semicircular
canals

ampullae of
semicircular canals

utricle

saccule

vestibule

vestibulocochlear nerve

organ of Corti
in cochlea

otoliths

gelatinous
material

hair
cells

nerve
fibers

c. Utricle and saccule

cochlea

tectorial
membrane

hair
cells

basilar
membrane

nerve
fibers

d. Organ of Corti in cochlea

Anatomy of the Inner Ear
Figure 19.14

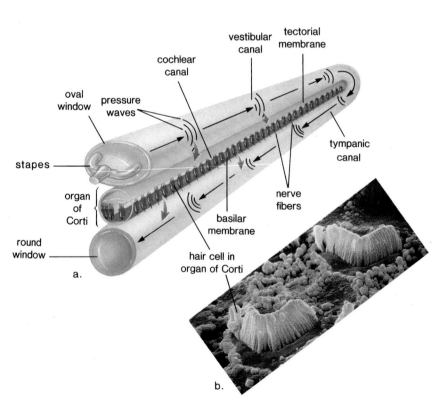

oval
window

pressure
waves

stapes

organ
of
Corti

round
window

a.

cochlear
canal

vestibular
canal

tectorial
membrane

tympanic
canal

nerve
fibers

basilar
membrane

hair cell in
organ of Corti

b.

Receptors for Hearing
Figure 19.16

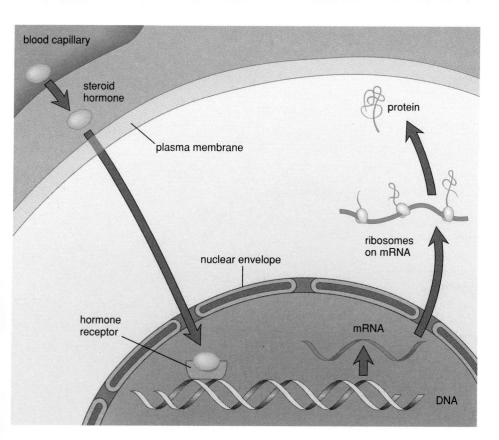

Cellular Activity of Steroid Hormones
Figure 20.1

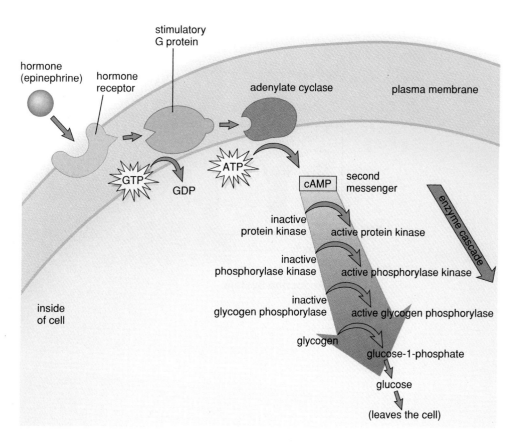

Cellular Activity of Peptide Hormones
Figure 20.2

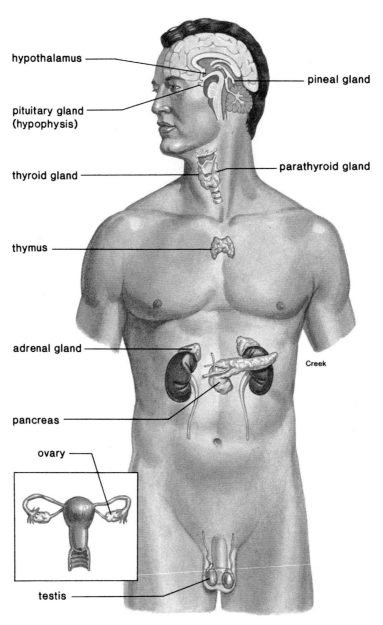

hypothalamus

pineal gland

pituitary gland
(hypophysis)

thyroid gland

parathyroid gland

thymus

adrenal gland

Creek

pancreas

ovary

testis

The Endocrine System
Figure 20.3

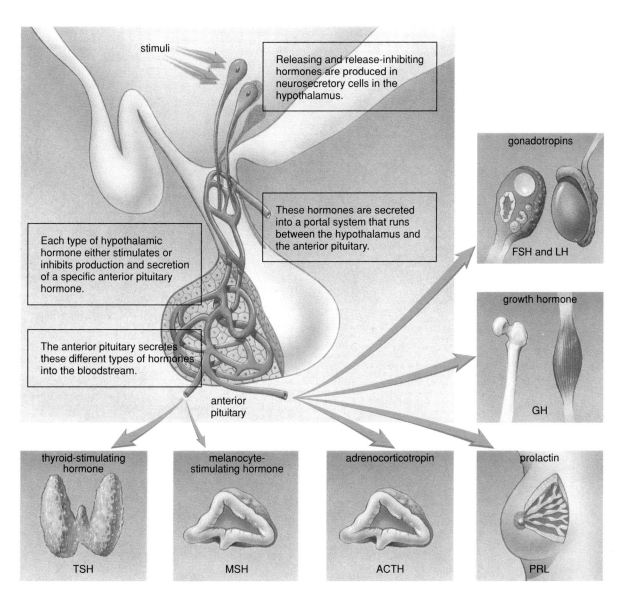

stimuli

Releasing and release-inhibiting hormones are produced in neurosecretory cells in the hypothalamus.

These hormones are secreted into a portal system that runs between the hypothalamus and the anterior pituitary.

Each type of hypothalamic hormone either stimulates or inhibits production and secretion of a specific anterior pituitary hormone.

The anterior pituitary secretes these different types of hormones into the bloodstream.

anterior pituitary

gonadotropins

FSH and LH

growth hormone

GH

thyroid-stimulating hormone

TSH

melanocyte-stimulating hormone

MSH

adrenocorticotropin

ACTH

prolactin

PRL

Hypothalamus and Anterior Pituitary
Figure 20.4

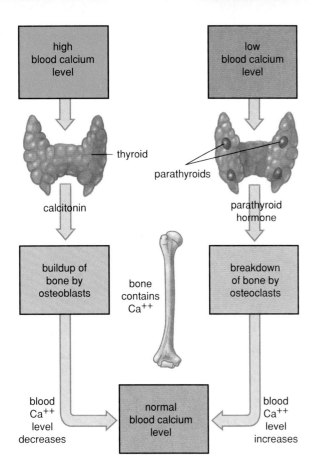

Regulation of Blood Calcium Level
Figure 20.9

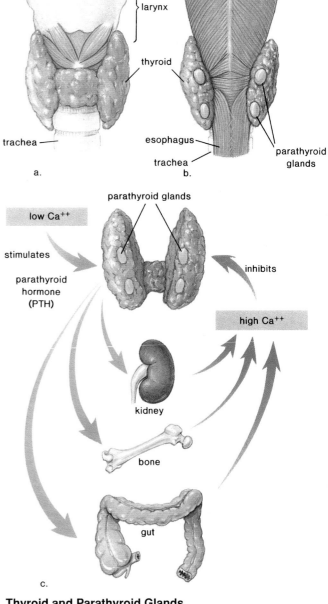

Thyroid and Parathyroid Glands
Figure 20.10

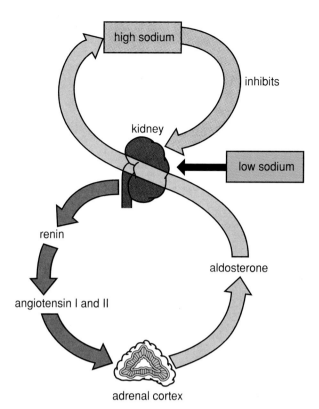

Renin-Angiotensin-Aldosterone Homeostatic System
Figure 20.12

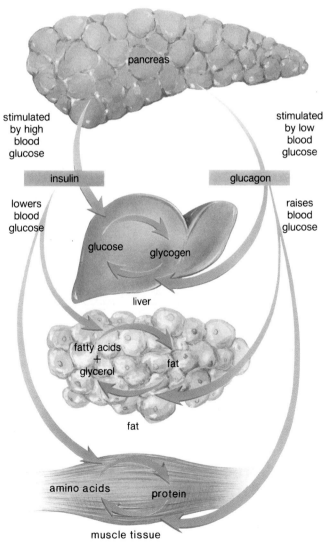

Insulin and Glucagon Homeostatic System
Figure 20.15

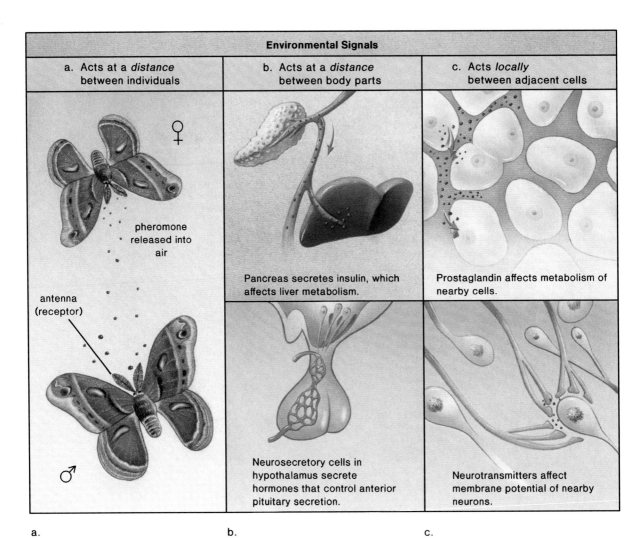

Environmental Signals		
a. Acts at a *distance* between individuals	b. Acts at a *distance* between body parts	c. Acts *locally* between adjacent cells

a. pheromone released into air

antenna (receptor)

♀

♂

b. Pancreas secretes insulin, which affects liver metabolism.

Neurosecretory cells in hypothalamus secrete hormones that control anterior pituitary secretion.

c. Prostaglandin affects metabolism of nearby cells.

Neurotransmitters affect membrane potential of nearby neurons.

a.

b.

c.

The Three Categories of Environmental Signals
Figure 20.16

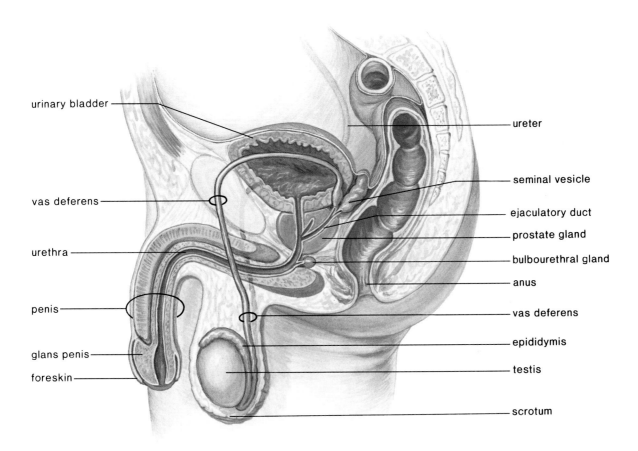

urinary bladder

ureter

vas deferens

seminal vesicle

ejaculatory duct

urethra

prostate gland

bulbourethral gland

anus

penis

vas deferens

epididymis

glans penis

testis

foreskin

scrotum

The Male Reproductive System
Figure 21.1

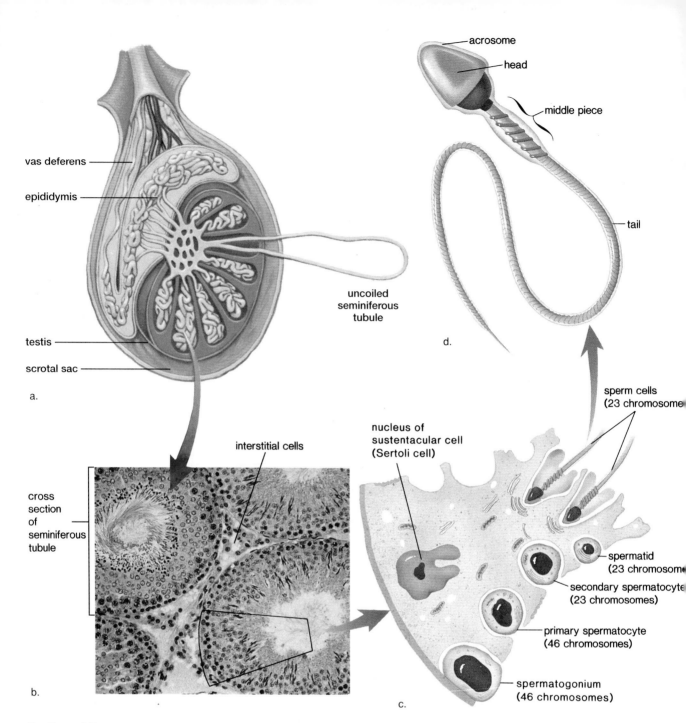

vas deferens

epididymis

testis

scrotal sac

a.

uncoiled
seminiferous
tubule

cross
section
of
seminiferous
tubule

interstitial cells

b.

acrosome

head

middle piece

tail

d.

sperm cells
(23 chromosome)

nucleus of
sustentacular cell
(Sertoli cell)

spermatid
(23 chromosome)

secondary spermatocyte
(23 chromosomes)

primary spermatocyte
(46 chromosomes)

spermatogonium
(46 chromosomes)

c.

Testis and Sperm
Figure 21.3

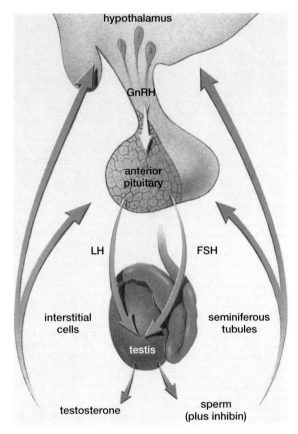

Hormonal Control of Testes
Figure 21.4

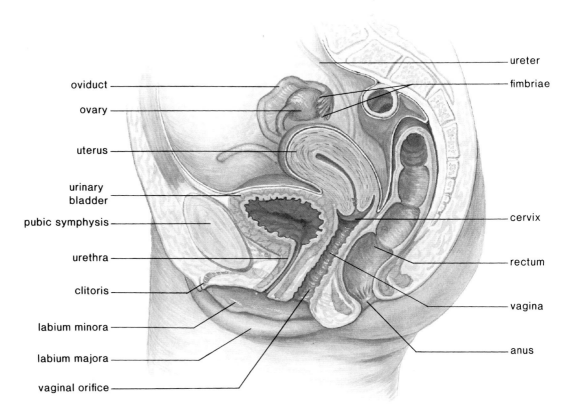

oviduct

ovary

uterus

urinary
bladder

pubic symphysis

urethra

clitoris

labium minora

labium majora

vaginal orifice

ureter

fimbriae

cervix

rectum

vagina

anus

The Female Reproductive System
Figure 21.5

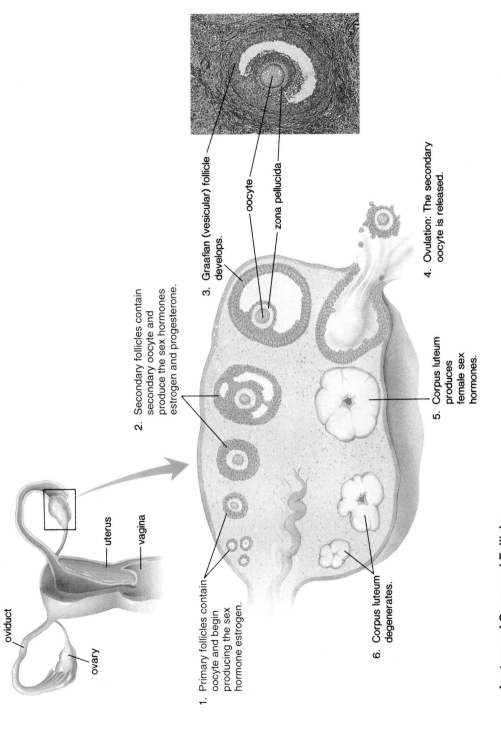

1. Primary follicles contain oocyte and begin producing the sex hormone estrogen.

2. Secondary follicles contain secondary oocyte and produce the sex hormones estrogen and progesterone.

3. Graafian (vesicular) follicle develops.

oocyte

zona pellucida

4. Ovulation: The secondary oocyte is released.

5. Corpus luteum produces female sex hormones.

6. Corpus luteum degenerates.

oviduct

ovary

uterus

vagina

Anatomy of Ovary and Follicle
Figure 21.7

The hypothalamus produces GnRH (gonadotropic-releasing hormone).

GnRH stimulates the anterior pituitary to produce FSH (follicle-stimulating hormone) and LH (luteinizing hormone).

FSH stimulates the follicle to produce estrogen and LH stimulates the corpus luteum to produce progesterone.

Estrogen and progesterone affect the sex organs (e.g., uterus) and the secondary sex characteristics and exert feedback control over the hypothalamus and the anterior pituitary.

Hormonal Control of Ovaries
Figure 21.8

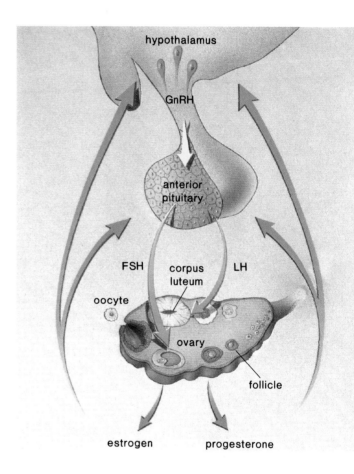

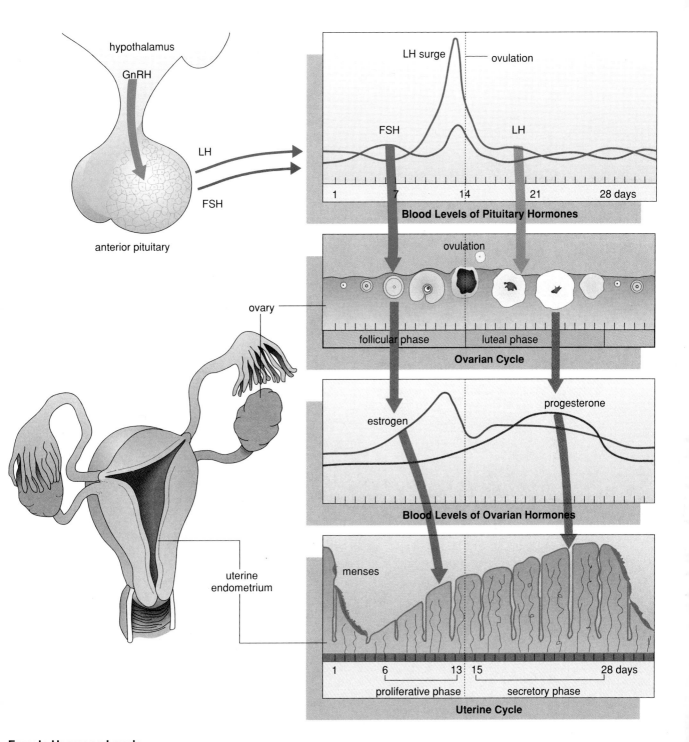

Female Hormone Levels
Figure 21.9

Within the figure, the following labels appear:

hypothalamus
GnRH
LH
FSH
anterior pituitary

LH surge — ovulation
FSH
LH
1   7   14   21   28 days
Blood Levels of Pituitary Hormones

ovulation
follicular phase   luteal phase
Ovarian Cycle

estrogen   progesterone
Blood Levels of Ovarian Hormones

ovary

uterine
endometrium

menses
1   6   13   15   28 days
proliferative phase   secretory phase
Uterine Cycle

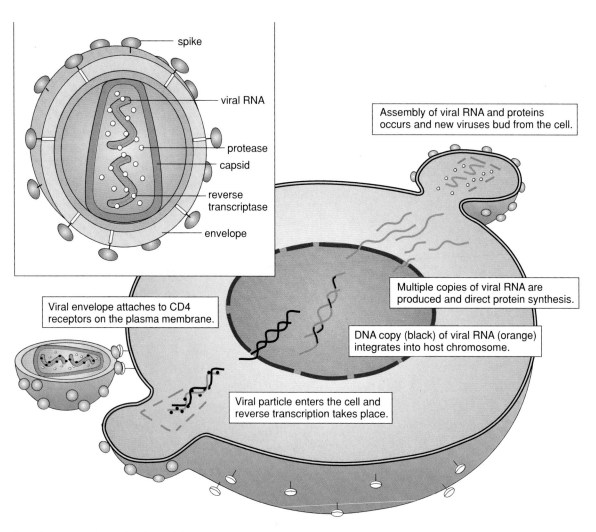

spike

viral RNA

protease

capsid

reverse transcriptase

envelope

Assembly of viral RNA and proteins occurs and new viruses bud from the cell.

Viral envelope attaches to CD4 receptors on the plasma membrane.

Multiple copies of viral RNA are produced and direct protein synthesis.

DNA copy (black) of viral RNA (orange) integrates into host chromosome.

Viral particle enters the cell and reverse transcription takes place.

Structure and Reproduction of HIV
Figure 21.14

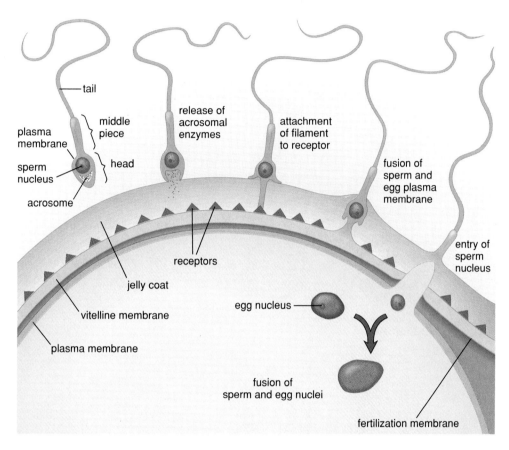

tail

release of
acrosomal
enzymes

attachment
of filament
to receptor

middle
piece

plasma
membrane

head

fusion of
sperm and
egg plasma
membrane

sperm
nucleus

acrosome

entry of
sperm
nucleus

receptors

jelly coat

egg nucleus

vitelline membrane

plasma membrane

fusion of
sperm and egg nuclei

fertilization membrane

Fertilization
Figure 22.1

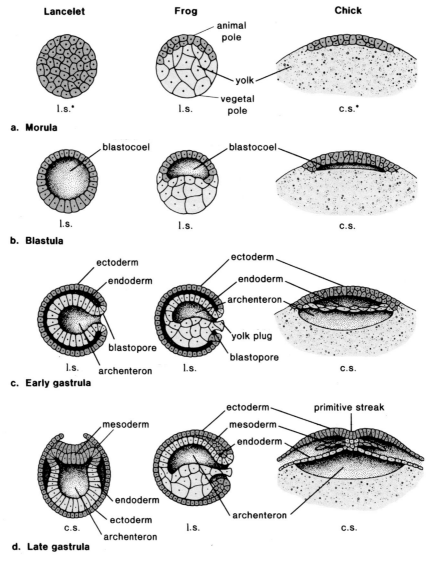

Lancelet   **Frog**   **Chick**

a. **Morula**

b. **Blastula**

c. **Early gastrula**

d. **Late gastrula**

*l.s. = longitudinal section; c.s. = cross section

Comparative Stages of Development
Figure 22.3

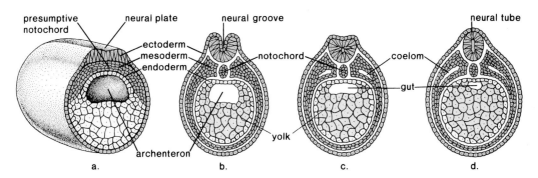

Development of Neural Tube and Coelom in a Frog Embryo
Figure 22.4

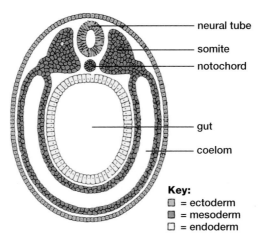

Key:
▨ = ectoderm
▨ = mesoderm
☐ = endoderm

Vertebrate Embryo Cross Section
Figure 22.5

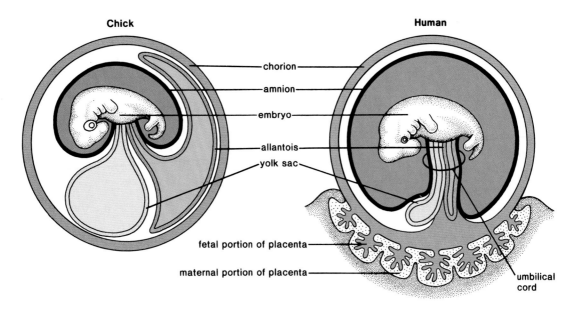

Chick   **Human**

chorion

amnion

embryo

allantois

yolk sac

fetal portion of placenta

maternal portion of placenta

umbilical cord

Extraembryonic Membranes
Figure 22.10

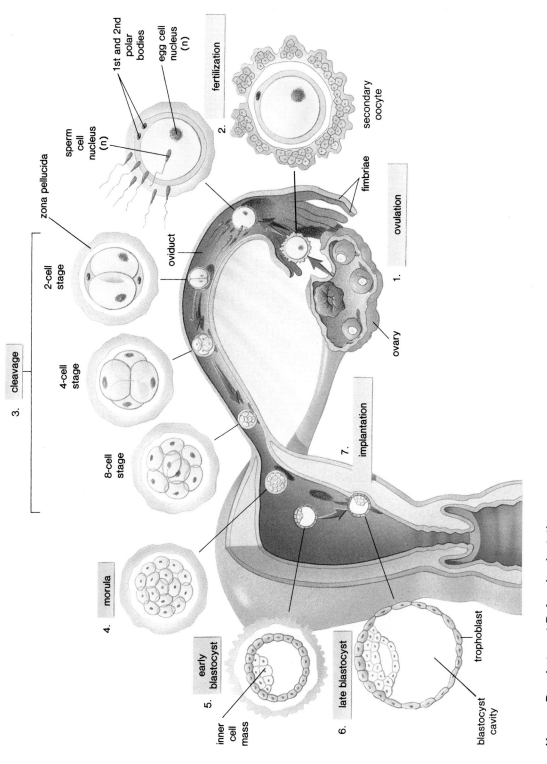

1st and 2nd polar bodies

egg cell nucleus (n)

sperm cell nucleus (n)

zona pellucida

fertilization

secondary oocyte

fimbriae

ovulation

oviduct

2-cell stage

4-cell stage

8-cell stage

cleavage

ovary

implantation

morula

early blastocyst

inner cell mass

late blastocyst

trophoblast

blastocyst cavity

1.
2.
3.
4.
5.
6.
7.

Human Development Before Implantation
Figure 22.11

137

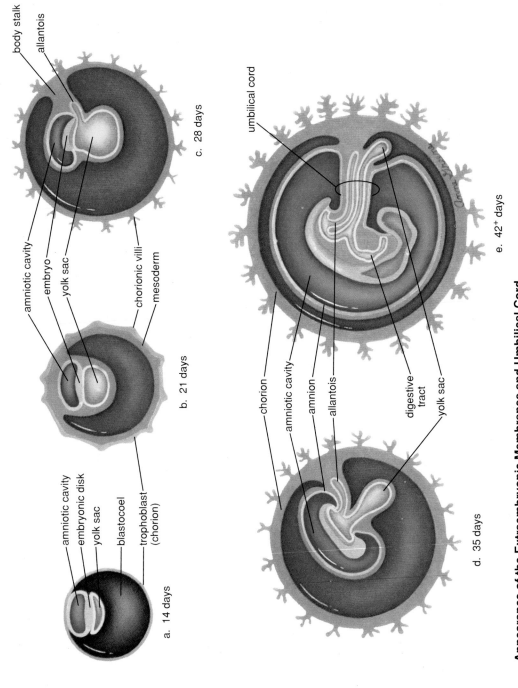

Appearance of the Extraembryonic Membranes and Umbilical Cord
Figure 22.12

a. 14 days

amniotic cavity
embryonic disk
yolk sac
blastocoel
trophoblast (chorion)

b. 21 days

amniotic cavity
embryo
yolk sac
chorionic villi
mesoderm

c. 28 days

body stalk
allantois

d. 35 days

e. 42⁺ days

chorion
amniotic cavity
amnion
allantois
digestive tract
yolk sac

umbilical cord

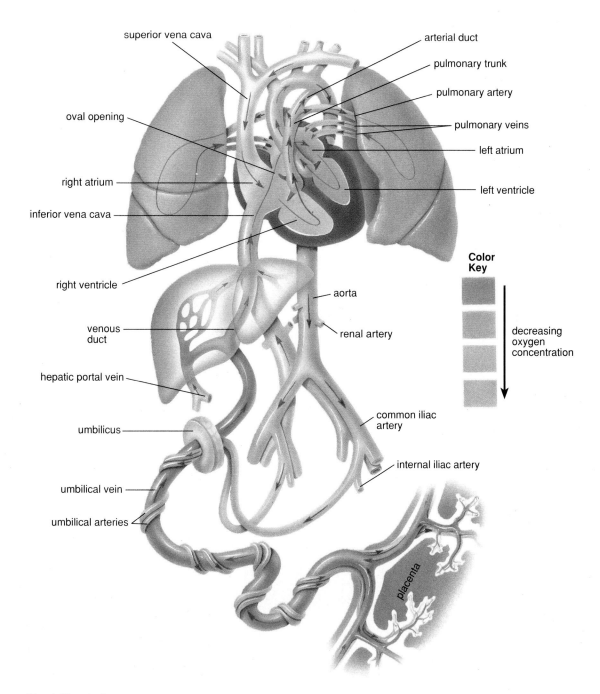

superior vena cava

arterial duct

pulmonary trunk

pulmonary artery

oval opening

pulmonary veins

left atrium

right atrium

left ventricle

inferior vena cava

right ventricle

aorta

venous duct

renal artery

hepatic portal vein

umbilicus

common iliac artery

internal iliac artery

umbilical vein

umbilical arteries

placenta

Color Key

decreasing oxygen concentration

Fetal Circulation
Figure 22.16

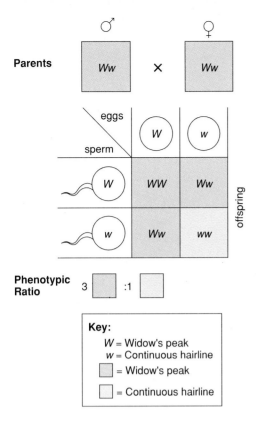

Parents

	♂		♀
	Ww	×	Ww

	eggs	
sperm	W	w
W	WW	Ww
w	Ww	ww

offspring

Phenotypic Ratio   3 ☐ :1 ☐

Key:
W = Widow's peak
w = Continuous hairline
☐ = Widow's peak
☐ = Continuous hairline

Monohybrid Testcross
Figure 23.4

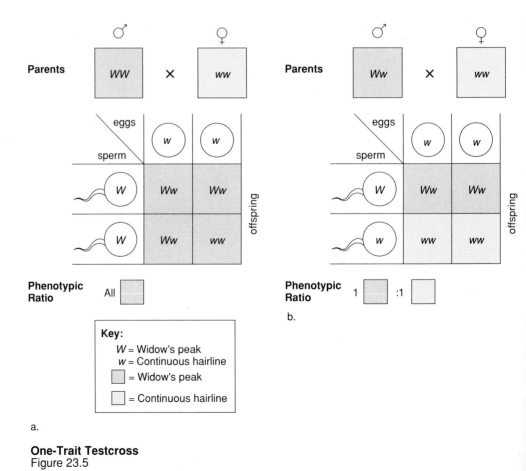

Parents

	♂		♀
	WW	×	ww

	eggs	
sperm	w	w
W	Ww	Ww
W	Ww	ww

offspring

Phenotypic Ratio   All ☐

a.

Parents

	♂		♀
	Ww	×	ww

	eggs	
sperm	w	w
W	Ww	Ww
w	ww	ww

offspring

Phenotypic Ratio   1 ☐ :1 ☐

b.

Key:
W = Widow's peak
w = Continuous hairline
☐ = Widow's peak
☐ = Continuous hairline

One-Trait Testcross
Figure 23.5

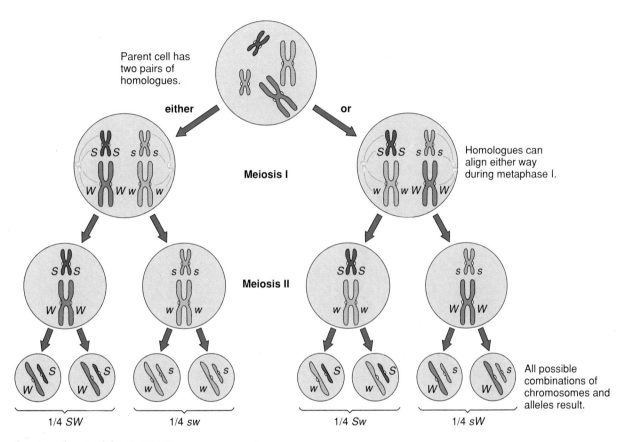

Parent cell has two pairs of homologues.

either ... **or**

Meiosis I

Homologues can align either way during metaphase I.

Meiosis II

All possible combinations of chromosomes and alleles result.

1/4 *SW*   1/4 *sw*   1/4 *Sw*   1/4 *sW*

Segregation and Assortment
Figure 23.6

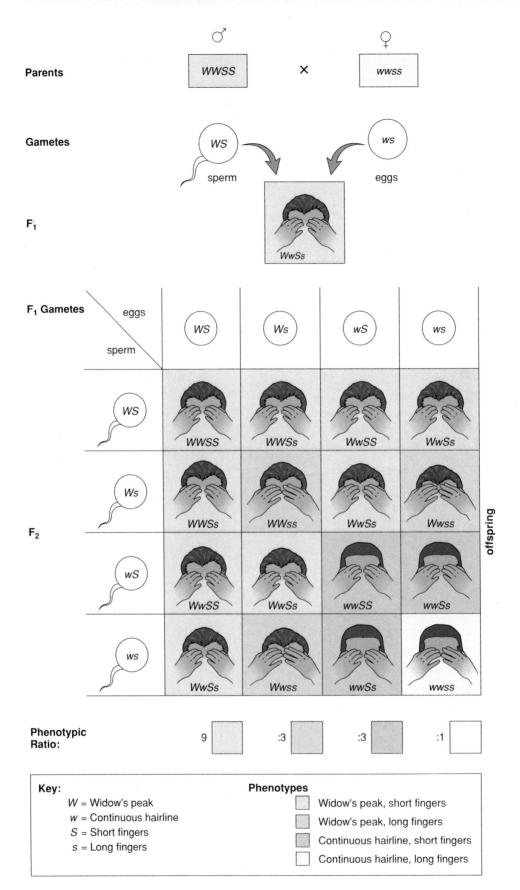

Parents	♂ WWSS	×	♀ wwss
Gametes	WS sperm		ws eggs
F₁	WwSs		

F₁ Gametes

eggs / sperm	WS	Ws	wS	ws
WS	WWSS	WWSs	WwSS	WwSs
Ws	WWSs	WWss	WwSs	Wwss
wS	WwSS	WwSs	wwSS	wwSs
ws	WwSs	Wwss	wwSs	wwss

F₂ — offspring

Phenotypic Ratio: 9 ☐ :3 ☐ :3 ☐ :1 ☐

Key:
W = Widow's peak
w = Continuous hairline
S = Short fingers
s = Long fingers

Phenotypes
☐ Widow's peak, short fingers
☐ Widow's peak, long fingers
☐ Continuous hairline, short fingers
☐ Continuous hairline, long fingers

Dihybrid Cross
Figure 23.7

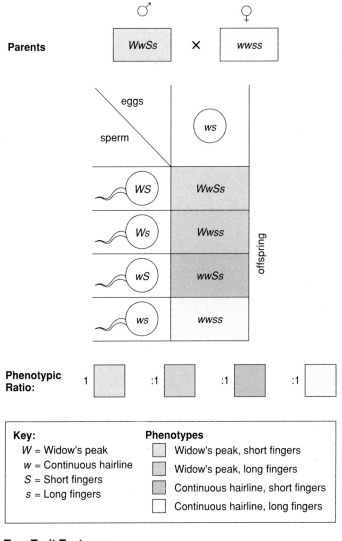

Parents

♂ WwSs × ♀ wwss

eggs

sperm

ws

WS   WwSs

Ws   Wwss

wS   wwSs

ws   wwss

offspring

Phenotypic Ratio: 1 [ ] :1 [ ] :1 [ ] :1 [ ]

Key:
W = Widow's peak
w = Continuous hairline
S = Short fingers
s = Long fingers

Phenotypes
[ ] Widow's peak, short fingers
[ ] Widow's peak, long fingers
[ ] Continuous hairline, short fingers
[ ] Continuous hairline, long fingers

Two-Trait Testcross
Figure 23.8

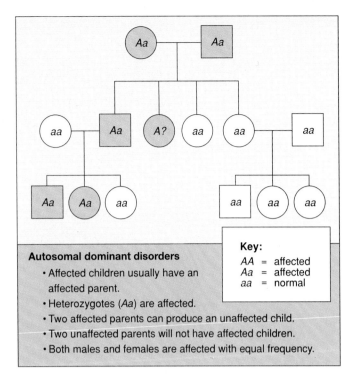

Autosomal dominant disorders

- Affected children usually have an affected parent.
- Heterozygotes (*Aa*) are affected.
- Two affected parents can produce an unaffected child.
- Two unaffected parents will not have affected children.
- Both males and females are affected with equal frequency.

Key:
AA = affected
Aa = affected
aa = normal

Autosomal Dominant Pedigree Chart
Figure 23.9

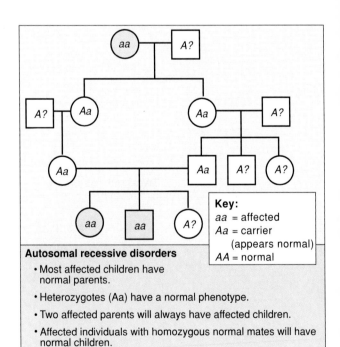

Autosomal recessive disorders

- Most affected children have normal parents.
- Heterozygotes (Aa) have a normal phenotype.
- Two affected parents will always have affected children.
- Affected individuals with homozygous normal mates will have normal children.
- Close relatives who reproduce are more likely to have affected children.
- Both males and females are affected with equal frequency.

Key:
aa = affected
Aa = carrier
(appears normal)
AA = normal

Autosomal Recessive Pedigree Chart
Figure 23.10

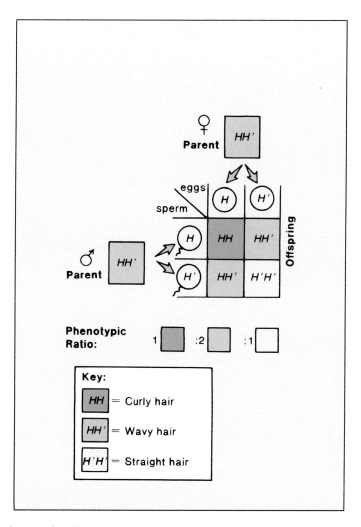

Incomplete Dominance
Figure 23.14

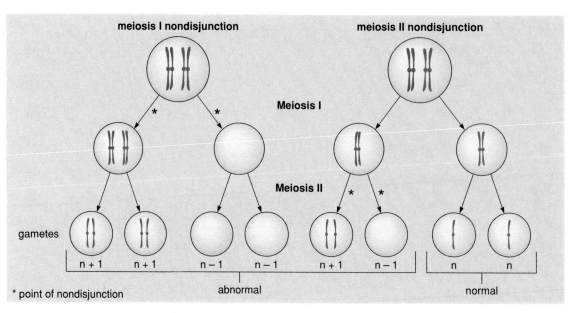

meiosis I nondisjunction

meiosis II nondisjunction

Meiosis I

Meiosis II

gametes

| n + 1 | n + 1 | n − 1 | n − 1 | n + 1 | n − 1 | n | n |

* point of nondisjunction

abnormal

normal

a.

Syndrome	Sex	Chromosomes	Frequency	
			Abortuses	*Births*
Down	M or F	Trisomy 21	1/40	1/700
Patau	M or F	Trisomy 13	1/33	1/15,000
Edward	M or F	Trisomy 18	1/200	1/6,000
Turner	F	XO	1/18	1/6,000
Metafemale	F	XXX (or XXXX)	0	1/1,500
Klinefelter	M	XXY (or XXXY)	0	1/1,500
Jacobs	M	XYY	?	1/1,000

b.

Nondisjunction of Autosomes
Figure 24.2

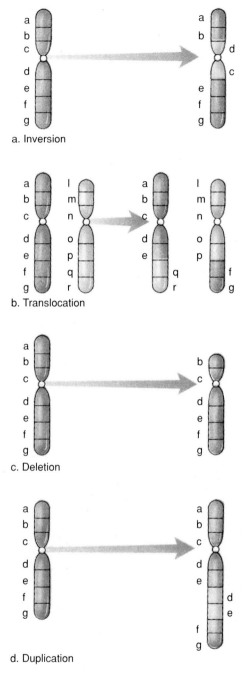

Types of Chromosome Mutations
Figure 24.4

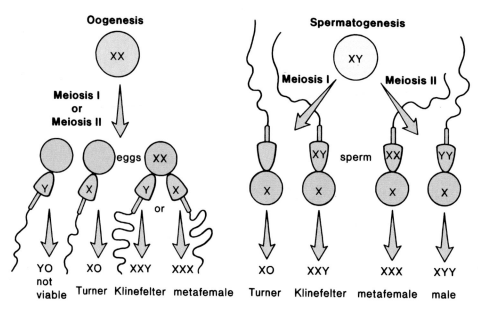

Nondisjunction of Sex Chromosomes
Figure 24.5

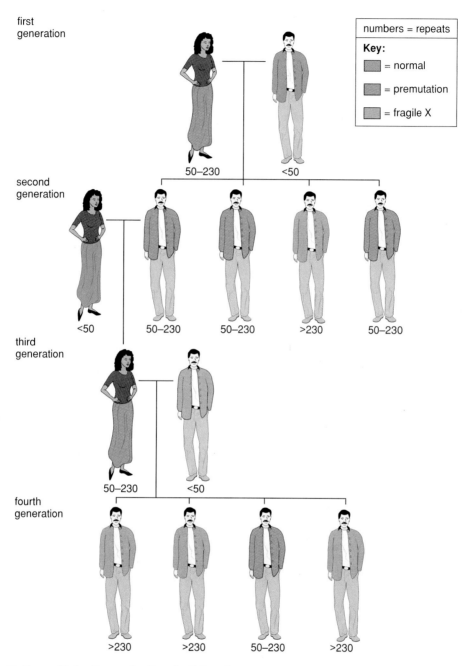

first
generation

second
generation

third
generation

fourth
generation

numbers = repeats
Key:
= normal
= premutation
= fragile X

50–230   <50

<50   50–230   50–230   >230   50–230

50–230   <50

>230   >230   50–230   >230

Pattern of Inheritance for Fragile X Syndrome
Figure 24A

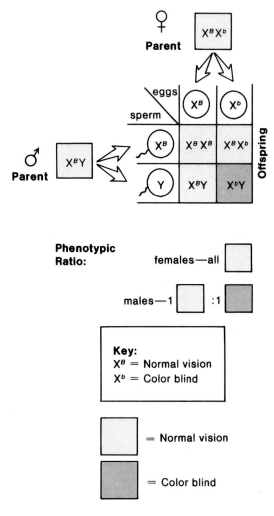

Phenotypic Ratio:

females—all

males—1   :1

Key:
X^B = Normal vision
X^b = Color blind

= Normal vision

= Color blind

Inheritance of a Sex-Linked Trait
Figure 24.8

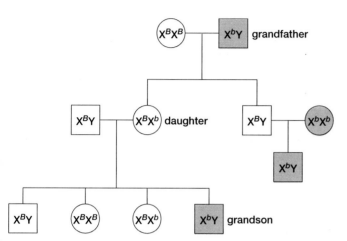

X-linked Recessive Disorders

Key:
X^BX^B = normal female X^BY = normal male
X^BX^b = carrier female X^bY = color-blind male
X^bX^b = color-blind female

a.

X-linked Recessive Disorders
Figure 24.9

X-linked Recessive Disorders

- More males than females are affected.
- An affected son can have parents who have the normal phenotype.
- For a female to have the characteristic, her father must also have it. Her mother must have it or be a carrier.
- The characteristic often skips a generation from the grandfather to the grandson.
- If a woman has the characteristic, all of her sons will have it.

b.

Among 205 catalogued X-linked recessive disorders are:

- Agammaglobulinemia—lack of immunity to infections
- Color blindness—inability to distinguish certain colors
- Hemophilia—defect in blood-clotting mechanisms
- Muscular dystrophy (some forms)—progressive wasting of muscles
- Spinal ataxia (some forms)—spinal cord degeneration

c.

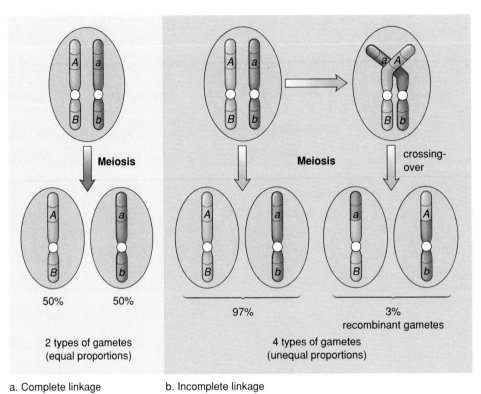

a. Complete linkage

b. Incomplete linkage

Linkage Group
Figure 24.12

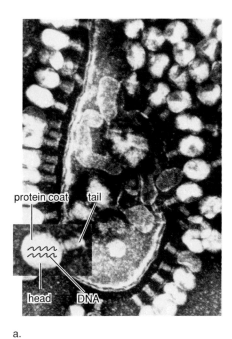

a.

Hershey and Chase Experiment
Figure 25.1

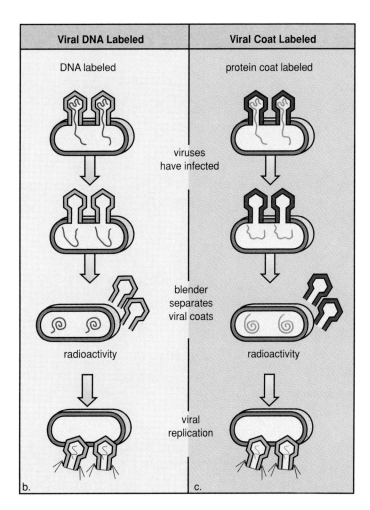

Viral DNA Labeled	Viral Coat Labeled
DNA labeled	protein coat labeled

viruses have infected

blender separates viral coats

radioactivity   radioactivity

viral replication

b.   c.

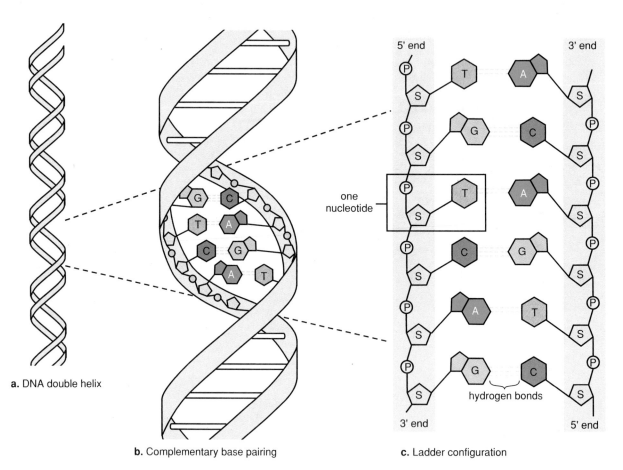

a. DNA double helix

b. Complementary base pairing

c. Ladder configuration

Overview of DNA Structure
Figure 25.2

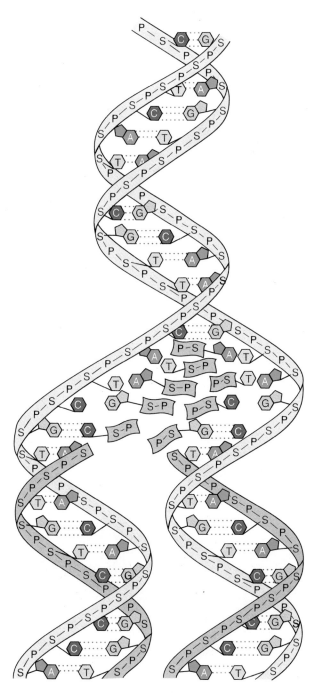

DNA Replication
Figure 25.4

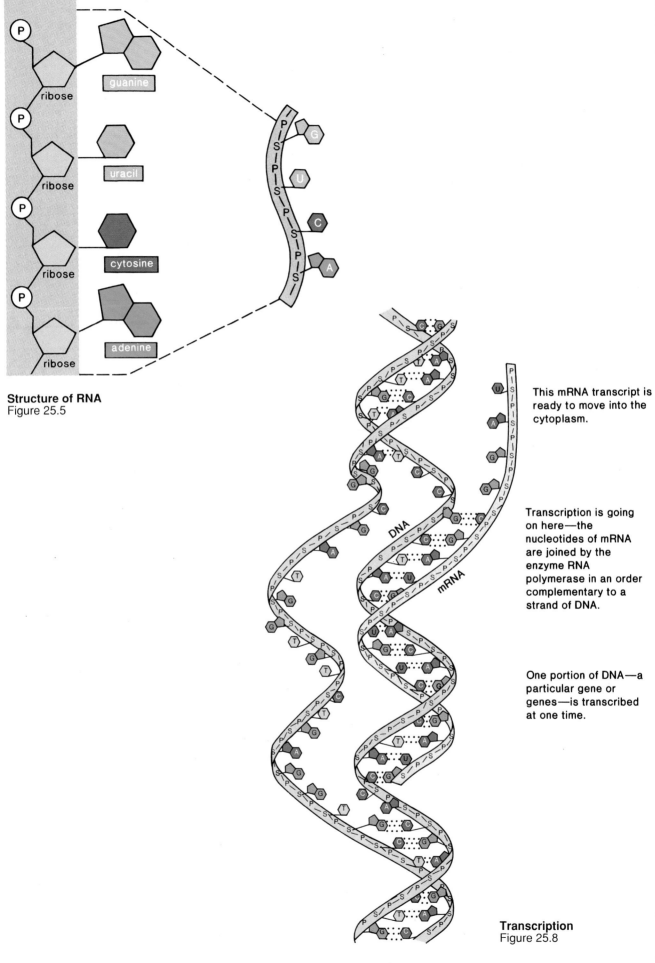

Structure of RNA
Figure 25.5

This mRNA transcript is ready to move into the cytoplasm.

Transcription is going on here—the nucleotides of mRNA are joined by the enzyme RNA polymerase in an order complementary to a strand of DNA.

One portion of DNA—a particular gene or genes—is transcribed at one time.

Transcription
Figure 25.8

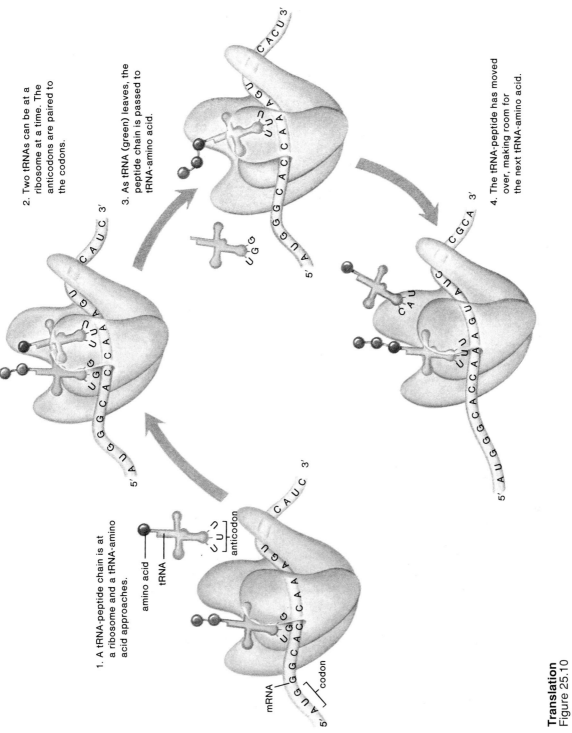

2. Two tRNAs can be at a ribosome at a time. The anticodons are paired to the codons.

3. As tRNA (green) leaves, the peptide chain is passed to tRNA-amino acid.

4. The tRNA-peptide has moved over, making room for the next tRNA-amino acid.

1. A tRNA-peptide chain is at a ribosome and a tRNA-amino acid approaches.

amino acid

tRNA

anticodon

mRNA

codon

Translation
Figure 25.10

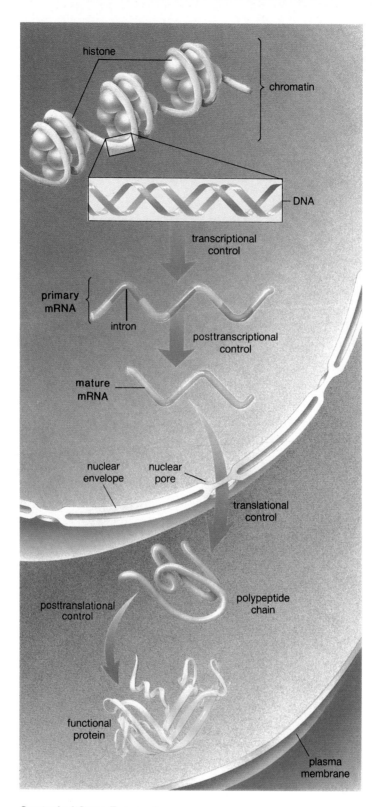

Control of Gene Expression
Figure 25.13

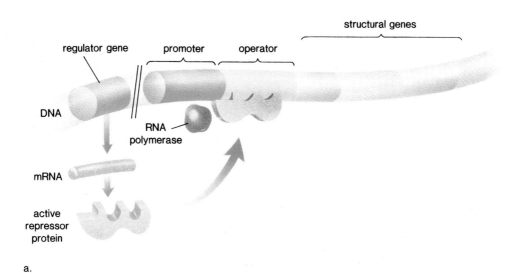

structural genes

regulator gene   promoter   operator

DNA

RNA
polymerase

mRNA

active
repressor
protein

a.

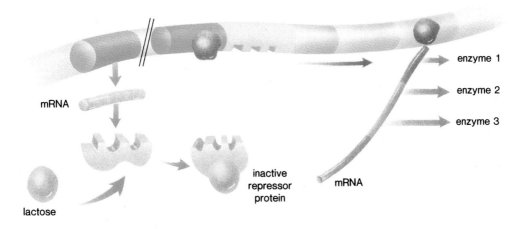

mRNA

enzyme 1

enzyme 2

enzyme 3

inactive
repressor
protein

mRNA

lactose

b.

The *lac* **Operon**
Figure 25.14

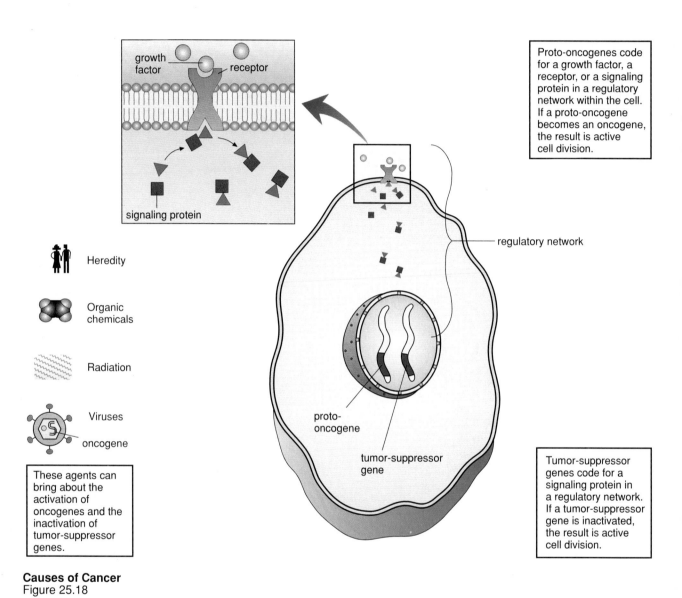

growth factor

receptor

signaling protein

Proto-oncogenes code for a growth factor, a receptor, or a signaling protein in a regulatory network within the cell. If a proto-oncogene becomes an oncogene, the result is active cell division.

regulatory network

Heredity

Organic chemicals

Radiation

Viruses

oncogene

These agents can bring about the activation of oncogenes and the inactivation of tumor-suppressor genes.

proto-oncogene

tumor-suppressor gene

Tumor-suppressor genes code for a signaling protein in a regulatory network. If a tumor-suppressor gene is inactivated, the result is active cell division.

Causes of Cancer
Figure 25.18

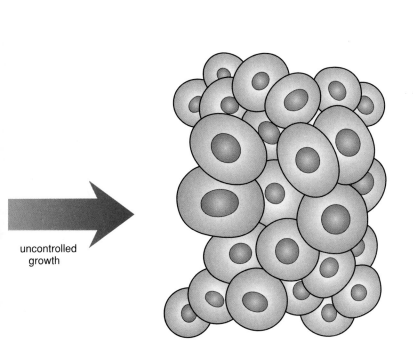

uncontrolled
growth

immune
system

Actively dividing
cancer cells have
all the character-
istics described
in Figure 25.17.

Some cancer cells
have altered plasma
membrane antigens
that subject them
to attack by the
lymphocytes of the
immune system.

Causes of Cancer
Figure 25.18 (cont.)

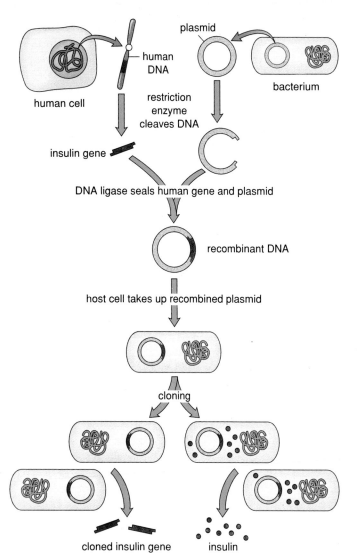

human cell

human DNA

plasmid

bacterium

restriction enzyme cleaves DNA

insulin gene

DNA ligase seals human gene and plasmid

recombinant DNA

host cell takes up recombined plasmid

cloning

cloned insulin gene

insulin

Cloning of a Human Gene
Figure 26.2

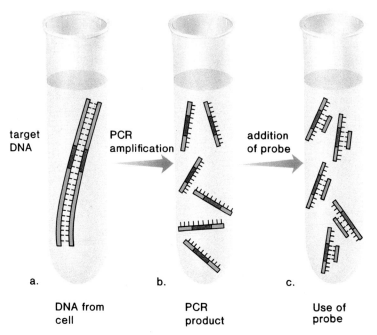

target DNA

PCR amplification

addition of probe

a.

b.

c.

DNA from cell

PCR product

Use of probe

Polymerase Chain Reaction (PCR)
Figure 26.3

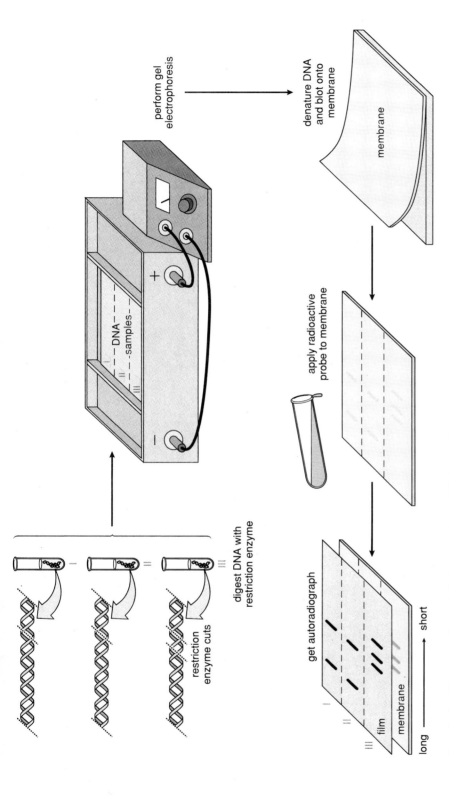

perform gel electrophoresis

denature DNA and blot onto membrane

apply radioactive probe to membrane

get autoradiograph

DNA samples

membrane

restriction enzyme cuts

digest DNA with restriction enzyme

long → short

film

membrane

Restriction Fragment Length Polymorphism (RFLP) Analysis
Figure 26.4

163

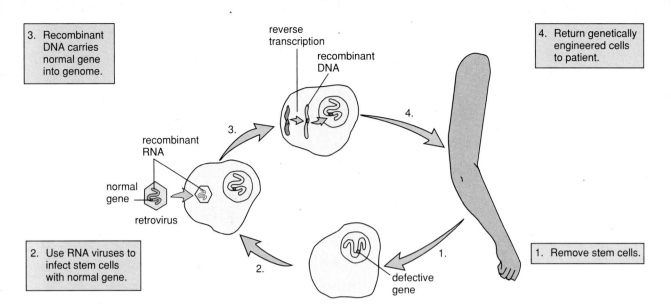

Ex vivo Gene Therapy in Humans
Figure 26.8

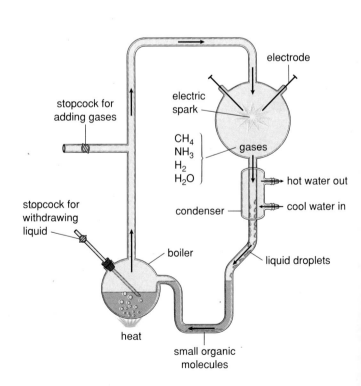

Miller's Experiment
Figure 27.1

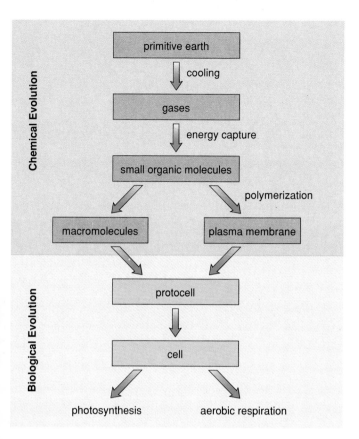

Chemical Evolution Produced the Protocell
Figure 27.3

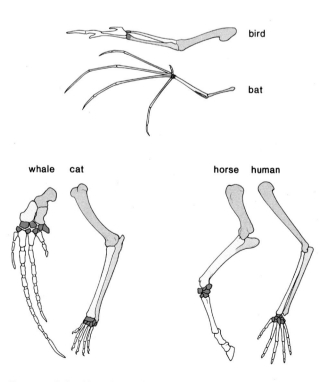

Bones of the Vertebrate Forelimbs
Figure 27.7

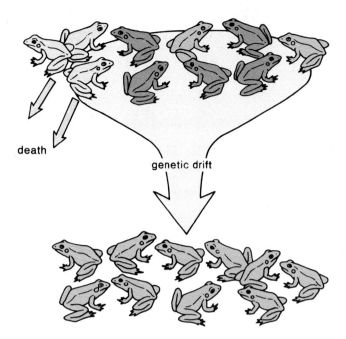

Genetic Drift
Figure 27.11

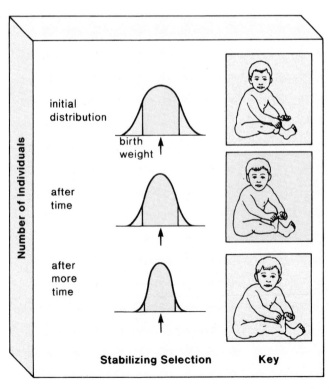

Stabilizing Selection
Figure 27.13

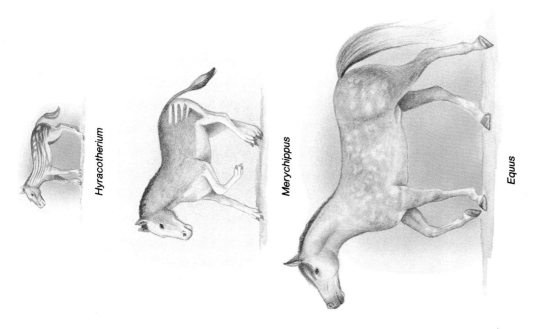

Hyracotherium

Merychippus

Equus

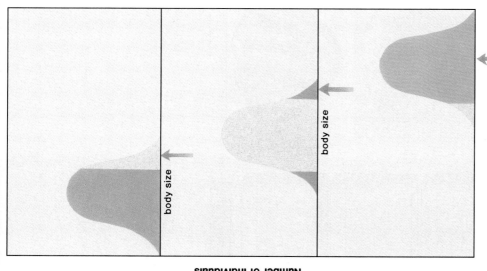

initial
distribution

after
time

after
more
time

Number of Individuals

body size

body size

Directional Selection

Directional Selection
Figure 27.14

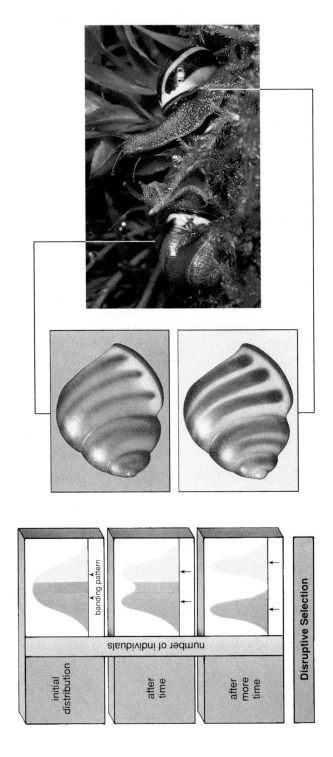

Disruptive Selection
Figure 27.15

168

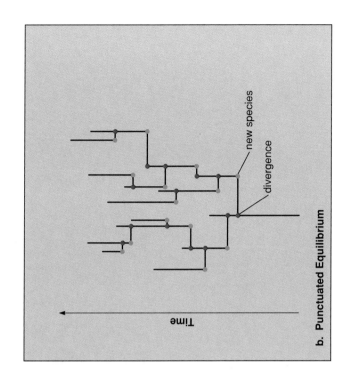

b. **Punctuated Equilibrium**

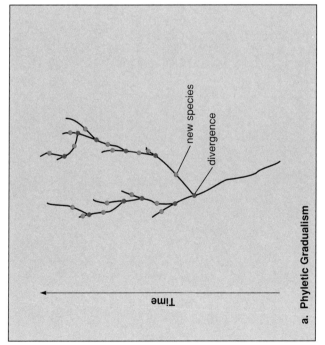

a. **Phyletic Gradualism**

Modes of Evolutionary Change
Figure 27.18

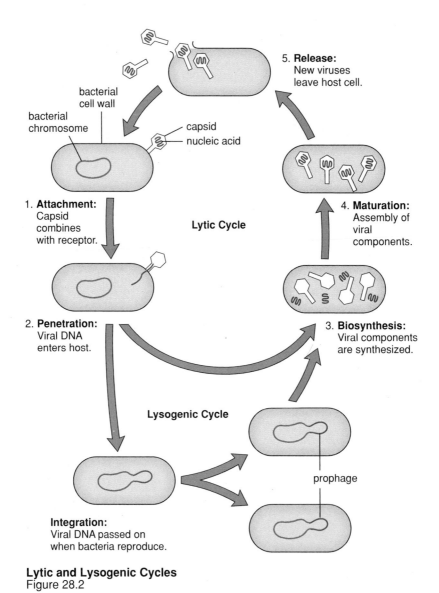

5. Release:
New viruses
leave host cell.

bacterial
cell wall

bacterial
chromosome

capsid
nucleic acid

1. Attachment:
Capsid
combines
with receptor.

Lytic Cycle

4. Maturation:
Assembly of
viral
components.

2. Penetration:
Viral DNA
enters host.

3. Biosynthesis:
Viral components
are synthesized.

Lysogenic Cycle

prophage

Integration:
Viral DNA passed on
when bacteria reproduce.

Lytic and Lysogenic Cycles
Figure 28.2

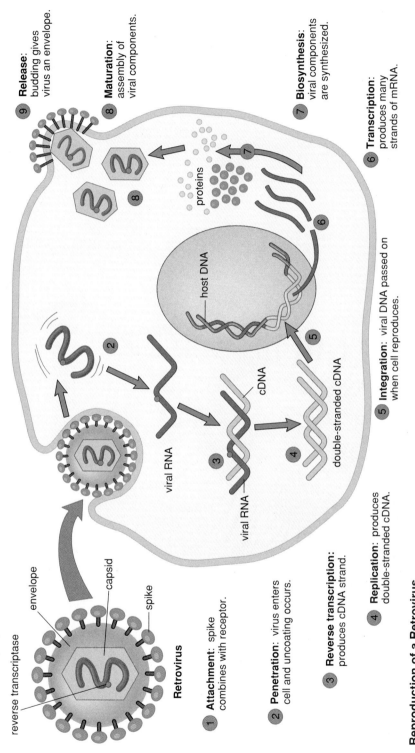

Retrovirus

reverse transcriptase
envelope
capsid
spike

① **Attachment:** spike combines with receptor.

② **Penetration:** virus enters cell and uncoating occurs.

③ **Reverse transcription:** produces cDNA strand.

④ **Replication:** produces double-stranded cDNA.

⑤ **Integration:** viral DNA passed on when cell reproduces.

⑥ **Transcription:** produces many strands of mRNA.

⑦ **Biosynthesis:** viral components are synthesized.

⑧ **Maturation:** assembly of viral components.

⑨ **Release:** budding gives virus an envelope.

viral RNA
viral RNA
cDNA
double-stranded cDNA
host DNA
proteins

Reproduction of a Retrovirus
Figure 28.3

171

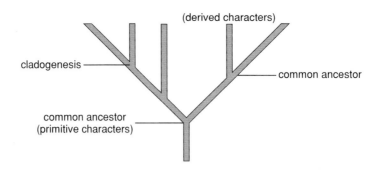

Evolutionary Tree
Figure 28A

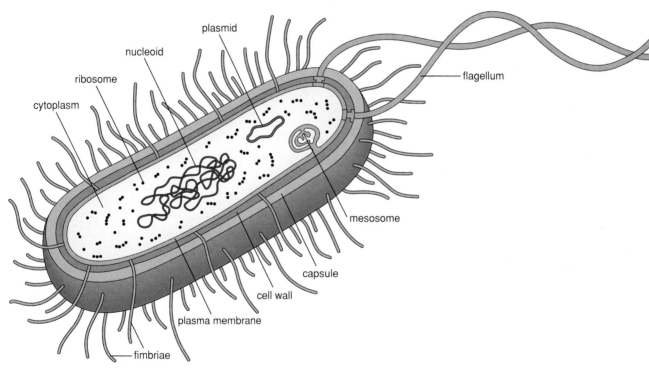

Typical Bacterial Cell
Figure 28.4

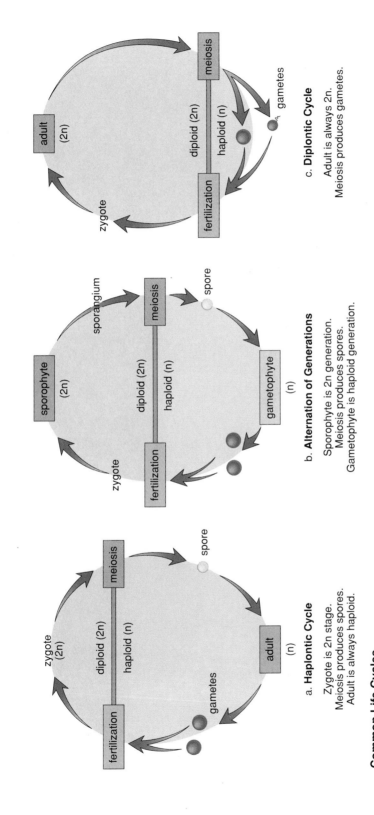

a. **Haplontic Cycle**

Zygote is 2n stage.
Meiosis produces spores.
Adult is always haploid.

b. **Alternation of Generations**

Sporophyte is 2n generation.
Meiosis produces spores.
Gametophyte is haploid generation.

c. **Diplontic Cycle**

Adult is always 2n.
Meiosis produces gametes.

Common Life Cycles
Figure 28.9

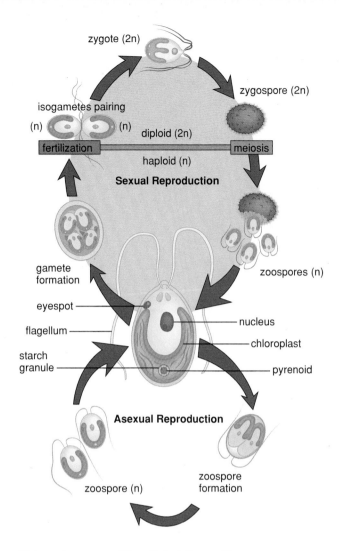

zygote (2n)

zygospore (2n)

isogametes pairing

(n) (n)

diploid (2n)

fertilization

meiosis

haploid (n)

Sexual Reproduction

gamete
formation

zoospores (n)

eyespot

nucleus

flagellum

chloroplast

starch
granule

pyrenoid

Asexual Reproduction

zoospore (n)

zoospore
formation

Chlamydomonas, **a Flagellated Green Alga**
Figure 28.10

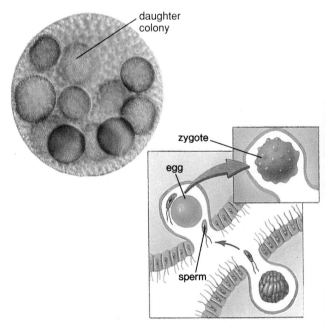

daughter
colony

zygote

egg

sperm

Volvox, **a Colonial Green Alga**
Figure 28.11

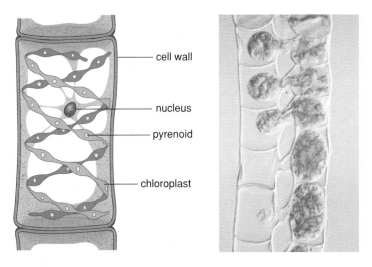

Spirogyra, a Filamentous Green Alga
Figure 28.12

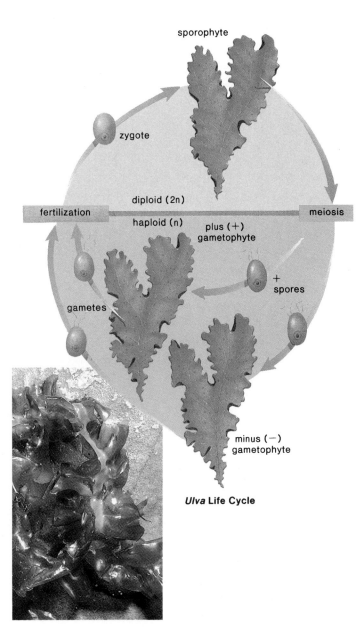

sporophyte

zygote

diploid (2n)

fertilization

haploid (n)

meiosis

plus (+)
gametophyte

+
spores

gametes

minus (−)
gametophyte

Ulva Life Cycle

Ulva, a Multicellular Green Alga
Figure 28.13

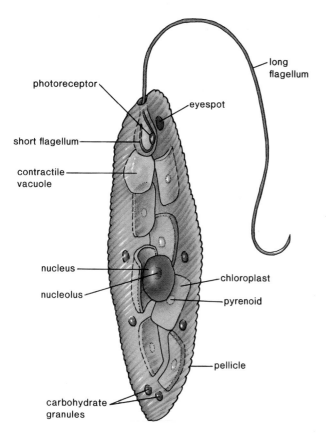

Euglena
Figure 28.16

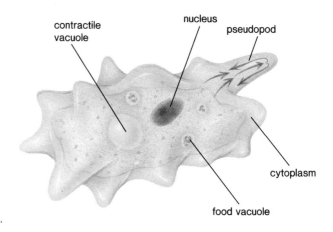

a.

b.

Amoeboid Protozoa
Figure 28.18

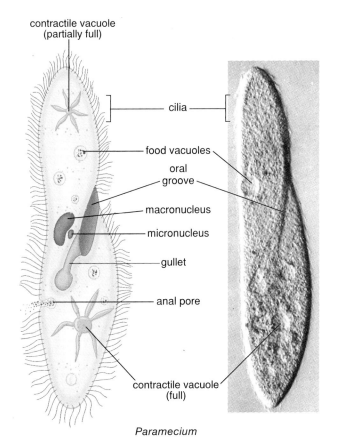

contractile vacuole
(partially full)

cilia

food vacuoles

oral
groove

macronucleus

micronucleus

gullet

anal pore

contractile vacuole
(full)

Paramecium

Ciliated Protozoa
Figure 28.19

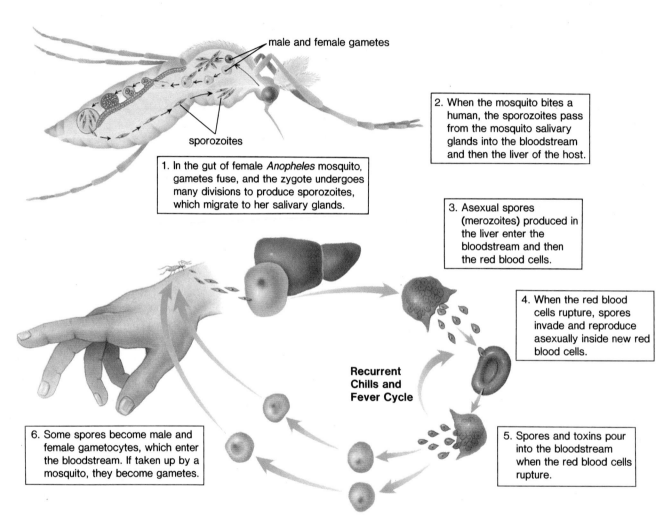

male and female gametes

2. When the mosquito bites a human, the sporozoites pass from the mosquito salivary glands into the bloodstream and then the liver of the host.

sporozoites

1. In the gut of female *Anopheles* mosquito, gametes fuse, and the zygote undergoes many divisions to produce sporozoites, which migrate to her salivary glands.

3. Asexual spores (merozoites) produced in the liver enter the bloodstream and then the red blood cells.

4. When the red blood cells rupture, spores invade and reproduce asexually inside new red blood cells.

Recurrent Chills and Fever Cycle

6. Some spores become male and female gametocytes, which enter the bloodstream. If taken up by a mosquito, they become gametes.

5. Spores and toxins pour into the bloodstream when the red blood cells rupture.

Life Cycle of *Plasmodium vivax*
Figure 28.21

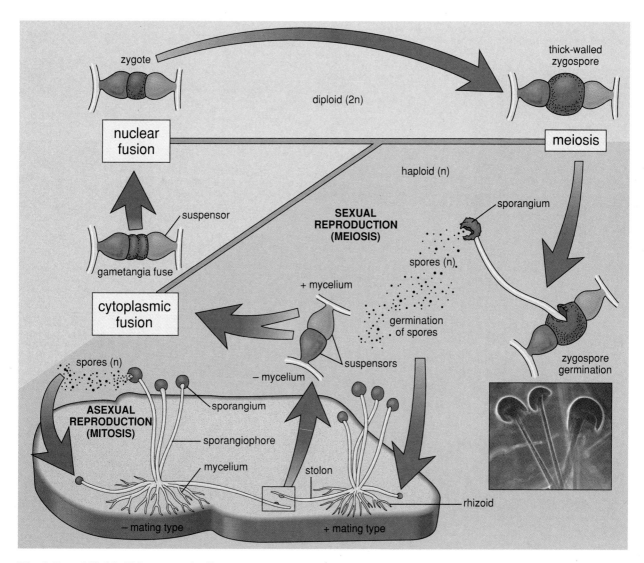

Black Bread Mold, *Rhizopus stolonifer*
Figure 28.23

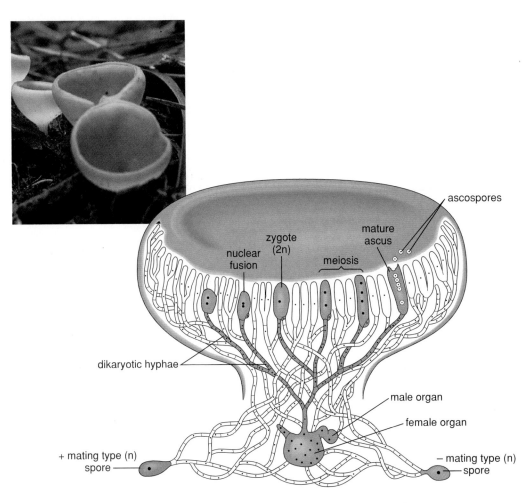

Sac Fungi
Figure 28.24a

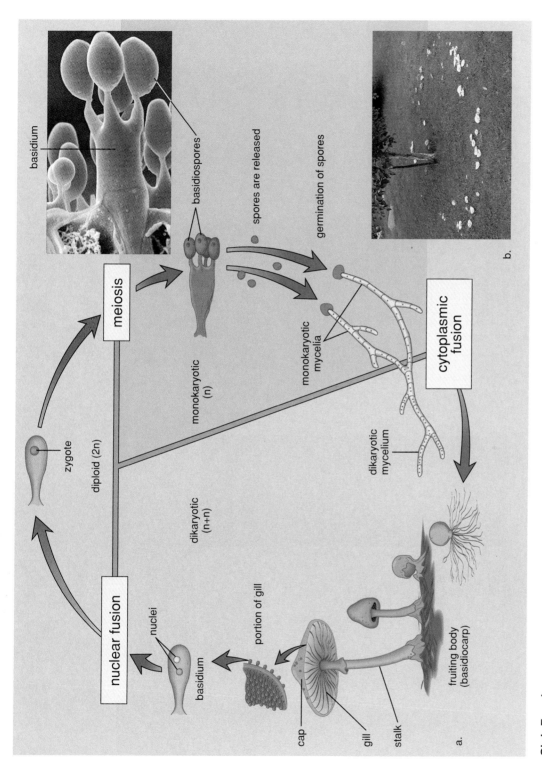

basidium

basidiospores

spores are released

germination of spores

meiosis

monokaryotic (n)

monokaryotic mycelia

cytoplasmic fusion

dikaryotic mycelium

zygote

diploid (2n)

dikaryotic (n+n)

nuclear fusion

nuclei

basidium

portion of gill

cap

gill

stalk

fruiting body (basidiocarp)

a.

b.

Club Fungi
Figure 28.25

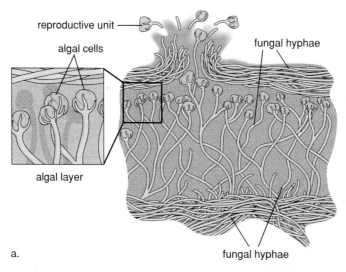

reproductive unit

algal cells

fungal hyphae

algal layer

fungal hyphae

a.

b.   Mixture of crustose lichens

Lichen Morphology
Figure 28.27

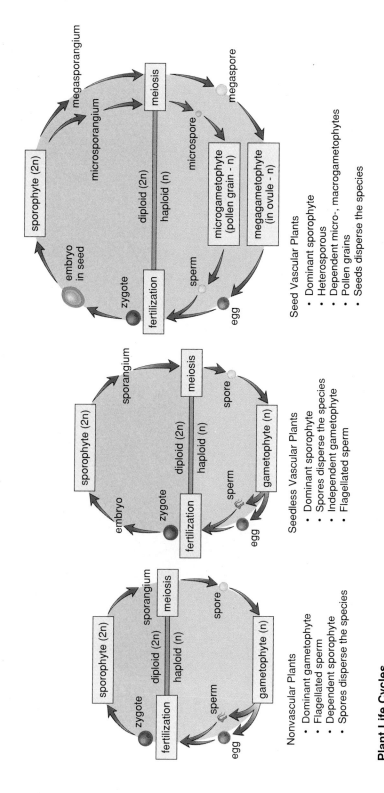

Nonvascular Plants
- Dominant gametophyte
- Flagellated sperm
- Dependent sporophyte
- Spores disperse the species

Seedless Vascular Plants
- Dominant sporophyte
- Spores disperse the species
- Independent gametophyte
- Flagellated sperm

Seed Vascular Plants
- Dominant sporophyte
- Heterosporous
- Dependent micro-, macrogametophytes
- Pollen grains
- Seeds disperse the species

Plant Life Cycles
Figure 29.2

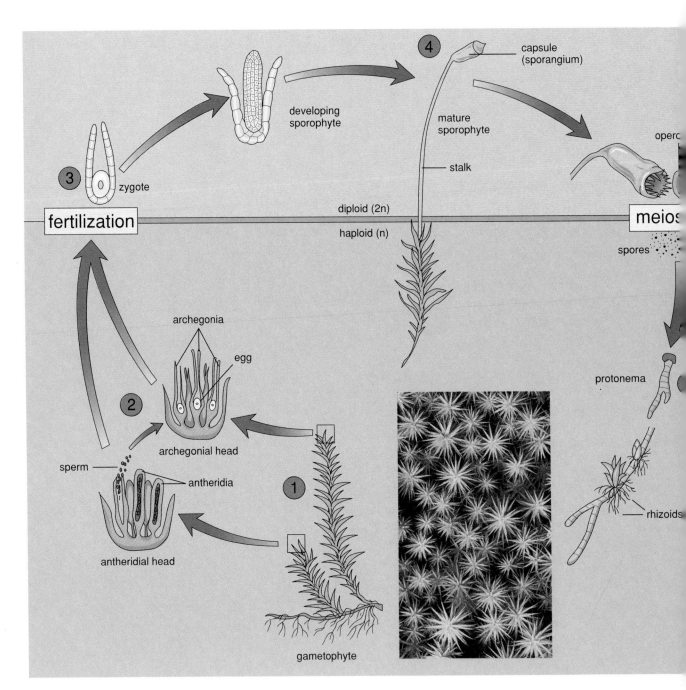

Moss Life Cycle
Figure 29.4

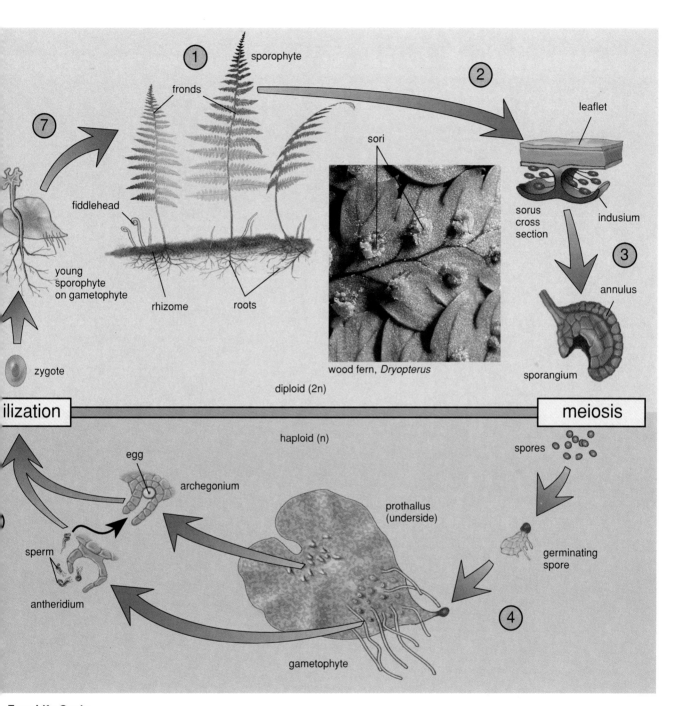

1 sporophyte

fronds

2 leaflet

7

fiddlehead

sori

sorus cross section

indusium

young sporophyte on gametophyte

rhizome

roots

wood fern, *Dryopterus*

3

annulus

sporangium

zygote

diploid (2n)

ilization   meiosis

haploid (n)

spores

egg

archegonium

prothallus (underside)

germinating spore

sperm

4

antheridium

gametophyte

Fern Life Cycle
Figure 29.10

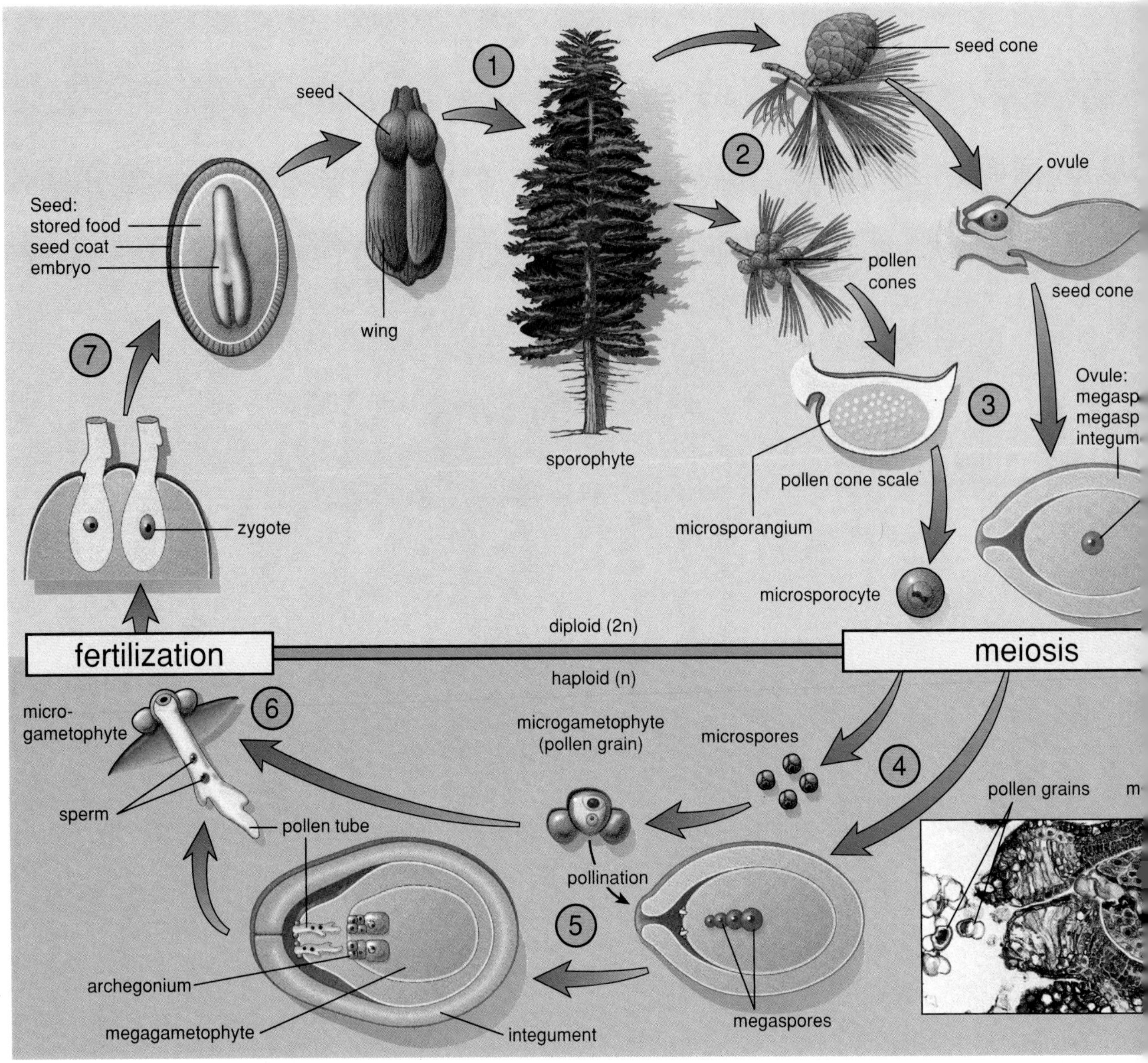
seed cone
1
seed
2
ovule
Seed:
stored food
seed coat
embryo
pollen
cones
seed cone
wing
7
Ovule:
megasp
megasp
integum
3
sporophyte
pollen cone scale
zygote
microsporangium
microsporocyte
diploid (2n)
fertilization
meiosis
haploid (n)
micro-
gametophyte
6
microgametophyte
(pollen grain)
microspores
4
pollen grains
m
sperm
pollen tube
pollination
5
archegonium
megagametophyte
integument
megaspores

Pine Life Cycle
Figure 29.12

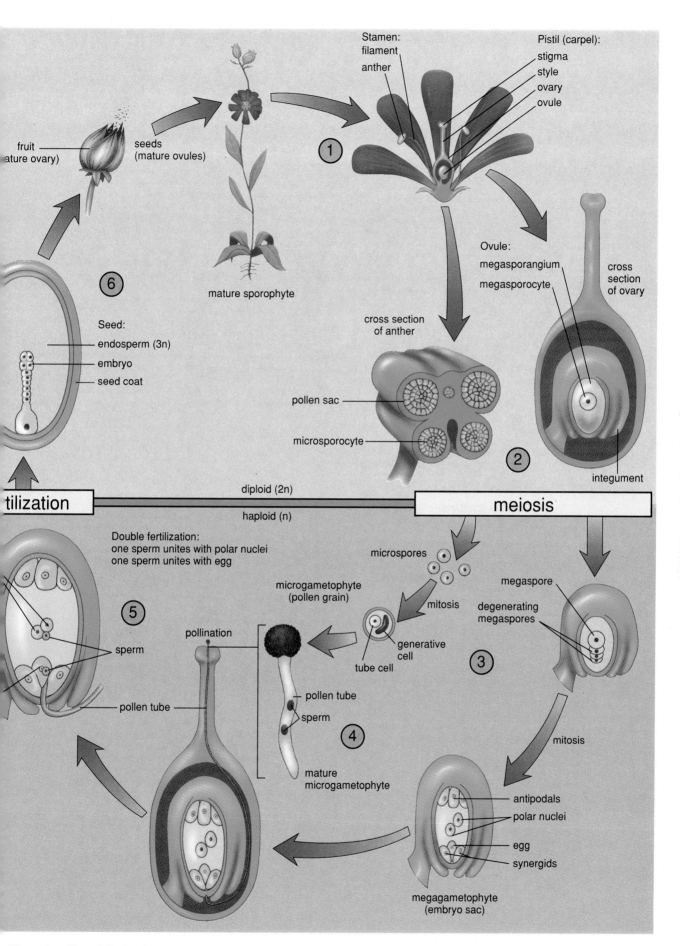

Flowering Plant Life Cycle
Figure 29.13

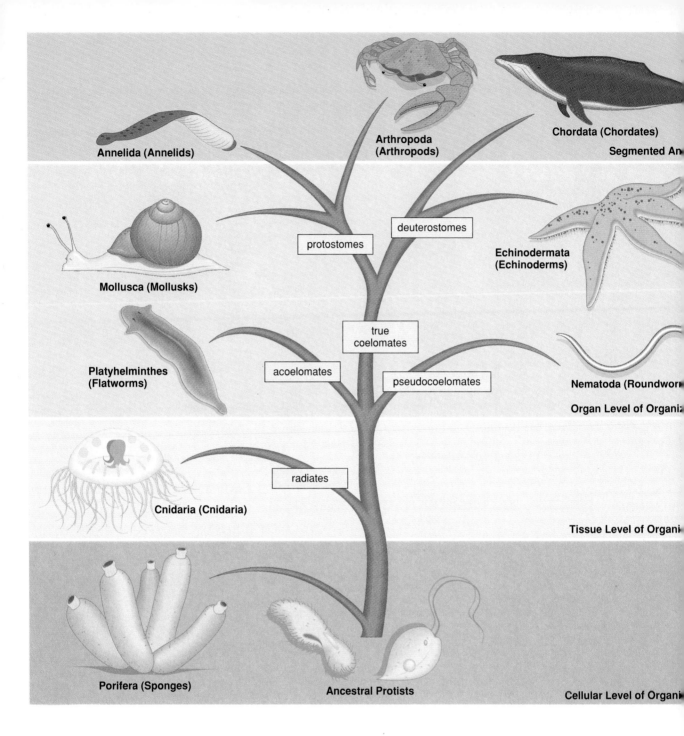

Evolutionary Tree For the Animal Kingdom
Figure 30.2

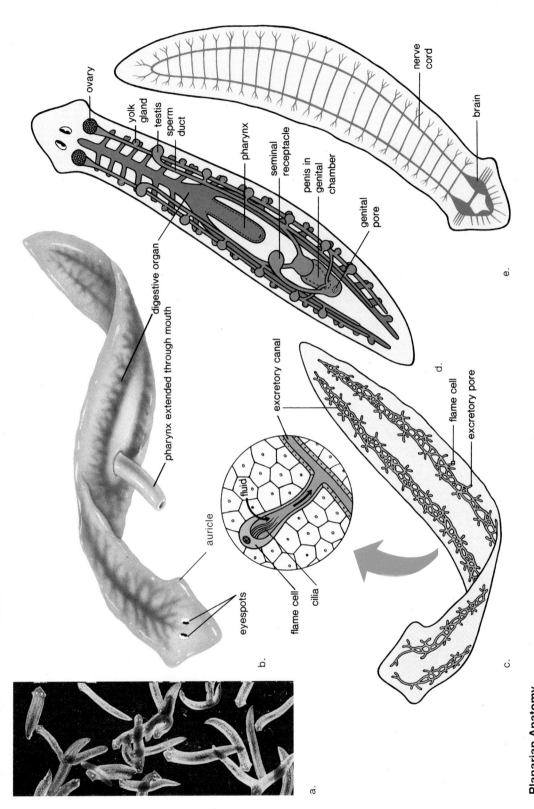

Planarian Anatomy
Figure 30.6

ovary

yolk gland

testis

sperm duct

pharynx

seminal receptacle

penis in genital chamber

genital pore

nerve cord

brain

e.

digestive organ

pharynx extended through mouth

auricle

eyespots

excretory canal

fluid

flame cell

cilia

b.

flame cell

excretory pore

d.

c.

a.

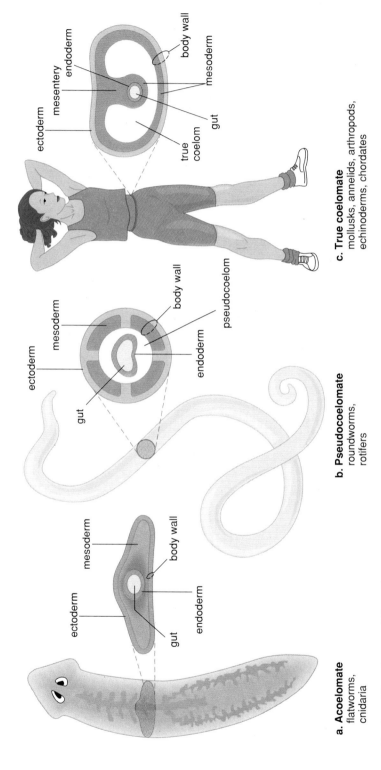

a. Acoelomate
flatworms,
cnidaria

ectoderm

mesoderm

body wall

gut

endoderm

b. Pseudocoelomate
roundworms,
rotifers

ectoderm

mesoderm

body wall

gut

endoderm

pseudocoelom

c. True coelomate
mollusks, annelids, arthropods,
echinoderms, chordates

ectoderm

mesentery

endoderm

body wall

mesoderm

gut

true
coelom

Coelom Structure and Function
Figure 30.8

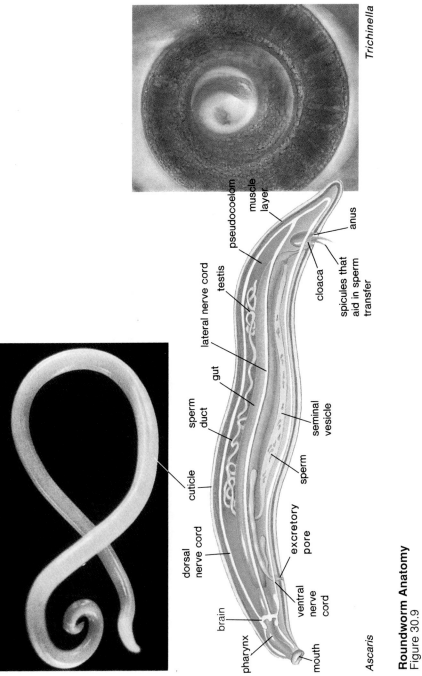

Trichinella

pseudocoelom

muscle layer

anus

lateral nerve cord

testis

cloaca

spicules that aid in sperm transfer

gut

sperm duct

seminal vesicle

cuticle

sperm

dorsal nerve cord

excretory pore

brain

ventral nerve cord

pharynx

mouth

Ascaris

Roundworm Anatomy
Figure 30.9

Protostomes

mollusks
annelids
arthropods

Deuterostomes

echinoderms
chordates

blastopore

mouth

blastopore

anus

primitive gut

anus

primitive gut

mouth

fate of blastopore

schizocoelom

enterocoelom

coelom forms by splitting of the mesoderm

coelom forms by out-pocketing of primitive gut

coelom formation

mouth

anus

trochophore larva

mouth

anus

dipleurula larva

larval form

Protostomes vs. Deuterostomes
Figure 30.10

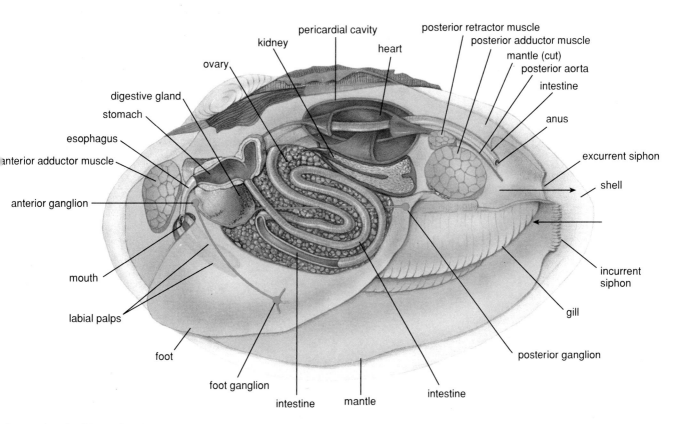

Anatomy of a Clam, *Anodonta*
Figure 30.12

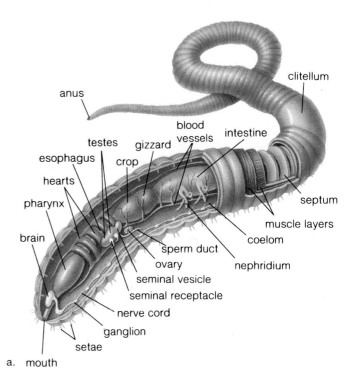

Anatomy of An Earthworm, *Lumbricus*
Figure 30.14a

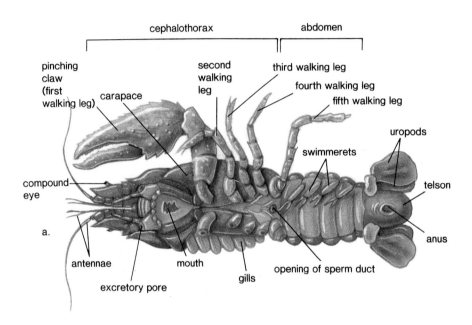

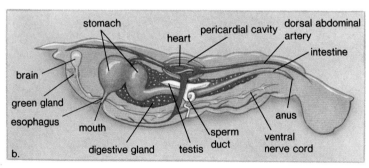

Anatomy of a Crayfish, *Homarus*
Figure 30.16

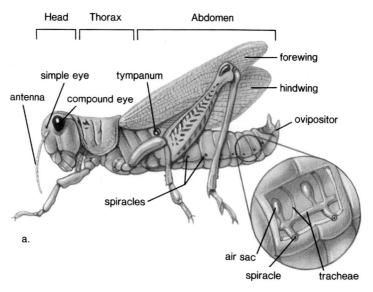

Head   Thorax   Abdomen

simple eye   tympanum   forewing

antenna   compound eye   hindwing

                                           ovipositor

spiracles

a.

                                           air sac

                                           spiracle   tracheae

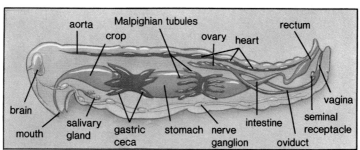

aorta   Malpighian tubules   rectum

crop   ovary   heart

brain   vagina

mouth   salivary   gastric   stomach nerve   intestine seminal
         gland   ceca   ganglion   receptacle
                                                  oviduct

Anatomy of a Female Grasshopper, *Romalea*
Figure 30.18

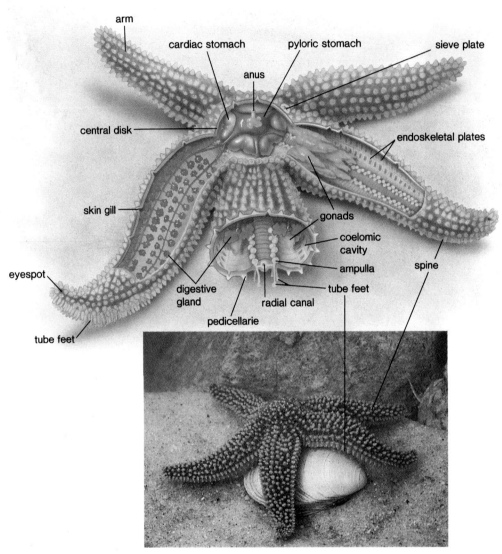

Starfish Anatomy
Figure 30.19b

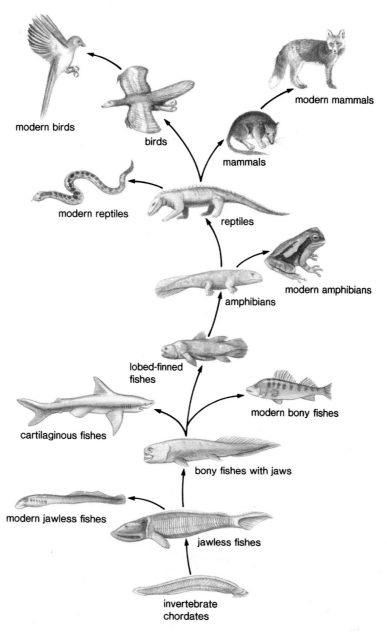

modern birds

birds

modern mammals

mammals

modern reptiles

reptiles

modern amphibians

amphibians

lobed-finned fishes

modern bony fishes

cartilaginous fishes

bony fishes with jaws

modern jawless fishes

jawless fishes

invertebrate chordates

Evolutionary Tree of Vertebrates
Figure 30.22

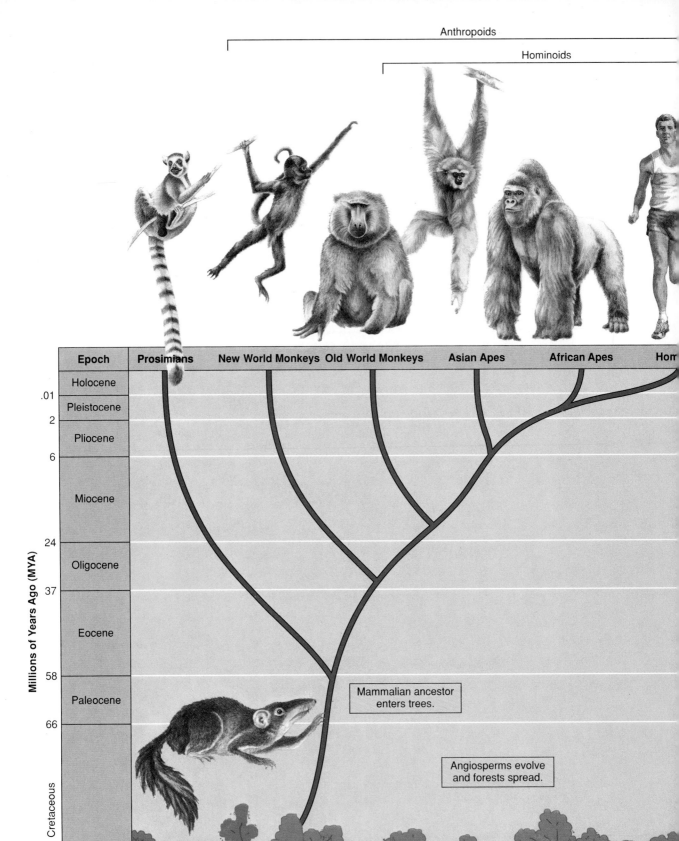

Primate Evolution
Figure 30.28

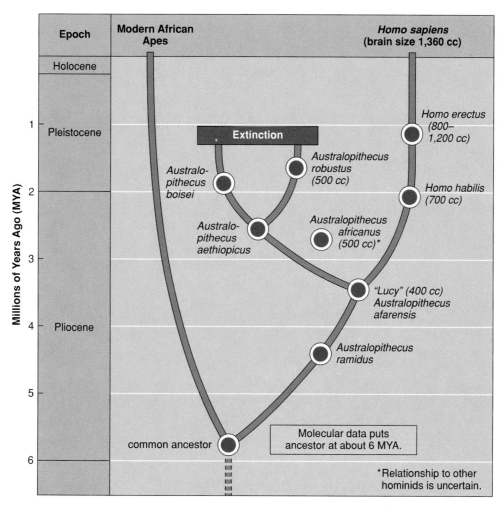

Recently Constructed Hominoid Evolutionary Tree
Figure 30.30

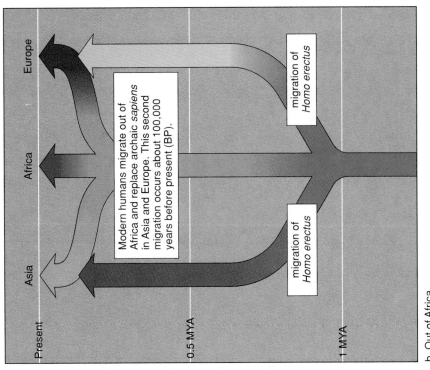

a. Multiregional continuity

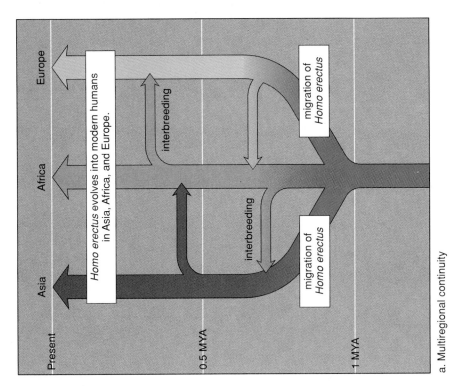

Modern humans migrate out of Africa and replace archaic *sapiens* in Asia and Europe. This second migration occurs about 100,000 years before present (BP).

b. Out of Africa

Two Hypotheses Concerning the Origins of Modern Humans
Figure 30.32

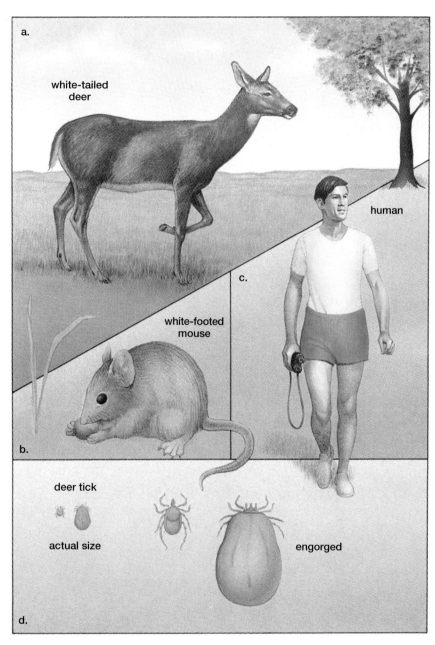

Life Cycle of a Deer Tick
Figure 32.7

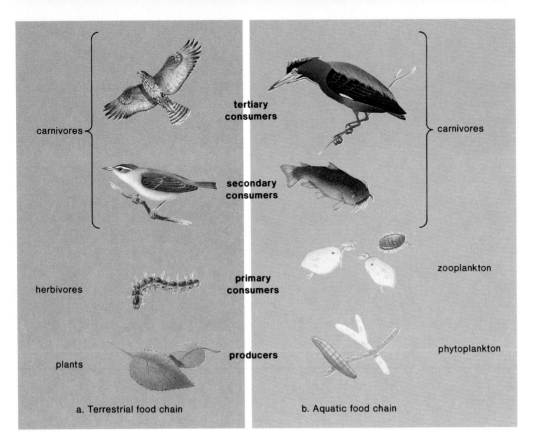

Examples of Food Chains
Figure 33.3

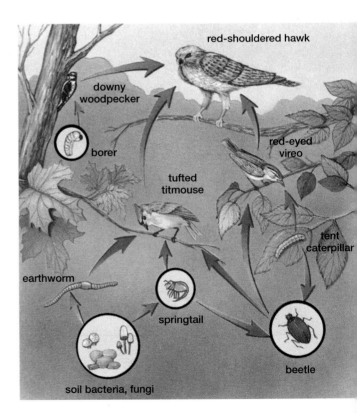

Deciduous Forest Ecosystem
Figure 33.4

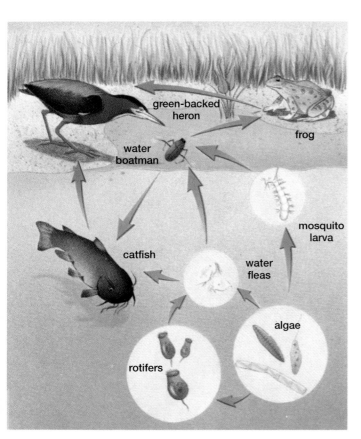

Freshwater Pond Ecosystem
Figure 33.5

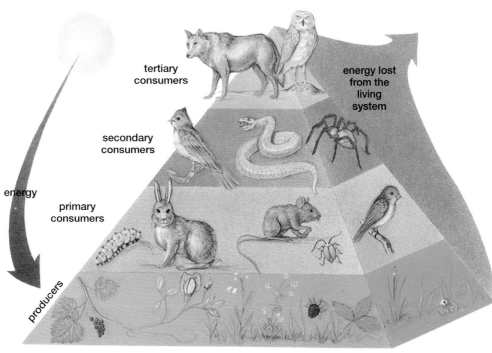

tertiary
consumers

energy lost
from the
living
system

secondary
consumers

energy

primary
consumers

producers

energy retained in the
living system

Pyramid of Energy
Figure 33.6

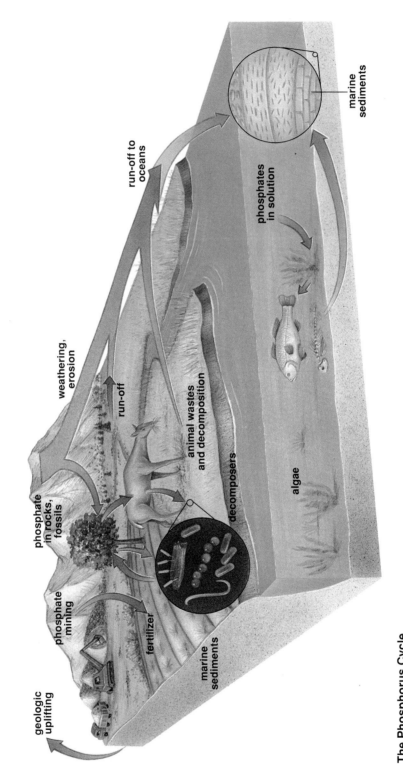

The Phosphorus Cycle
Figure 33.8

geologic
uplifting

phosphate
mining

phosphate
in rocks,
fossils

weathering,
erosion

run-off

fertilizer

marine
sediments

animal wastes
and decomposition

decomposers

algae

run-off to
oceans

phosphates
in solution

marine
sediments

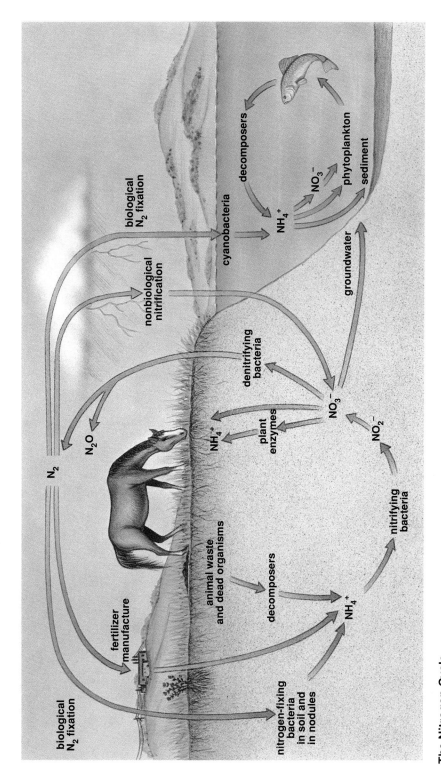

The Nitrogen Cycle
Figure 33.9

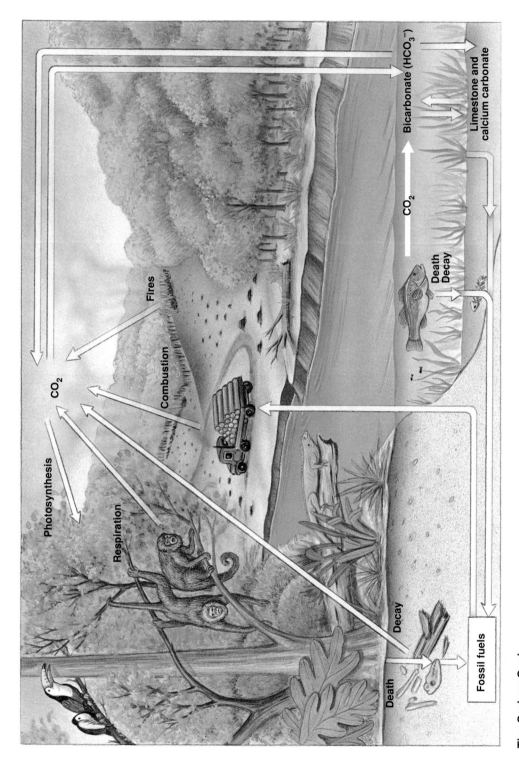

The Carbon Cycle
Figure 33.10

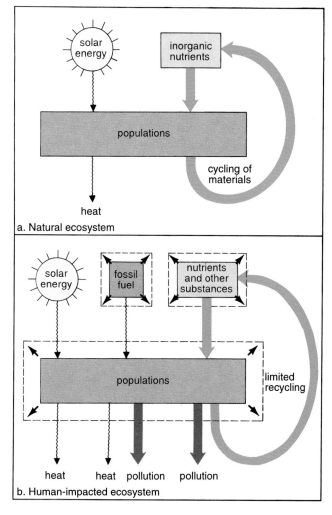

The Natural System vs. the Human-Impacted System
Figure 33.11

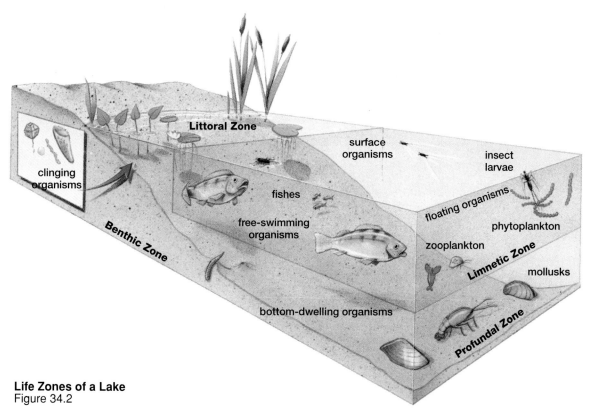

Life Zones of a Lake
Figure 34.2

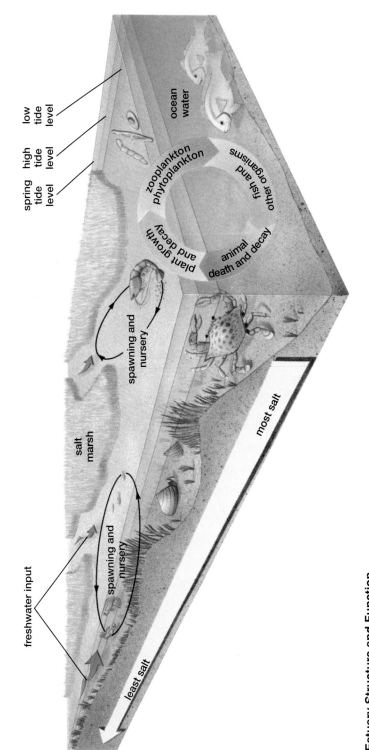

Estuary Structure and Function
Figure 34.3

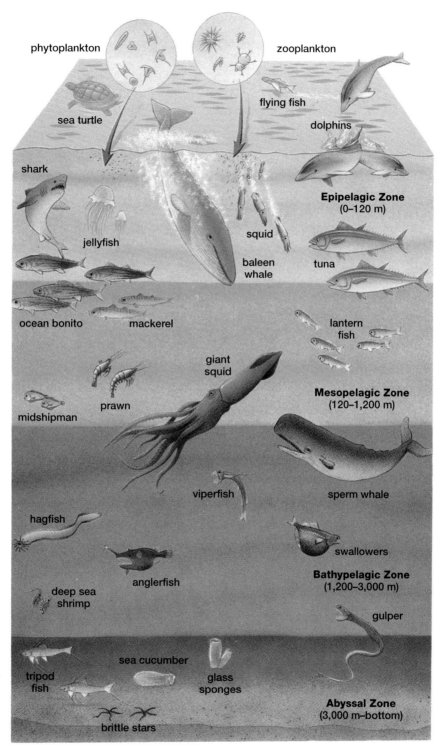

phytoplankton

zooplankton

sea turtle

flying fish

dolphins

shark

Epipelagic Zone
(0–120 m)

jellyfish

squid

baleen
whale

tuna

ocean bonito

mackerel

lantern
fish

giant
squid

Mesopelagic Zone
(120–1,200 m)

prawn

midshipman

viperfish

sperm whale

hagfish

swallowers

deep sea
shrimp

anglerfish

Bathypelagic Zone
(1,200–3,000 m)

gulper

sea cucumber

tripod
fish

glass
sponges

Abyssal Zone
(3,000 m–bottom)

brittle stars

Pelagic Division
Figure 34.7

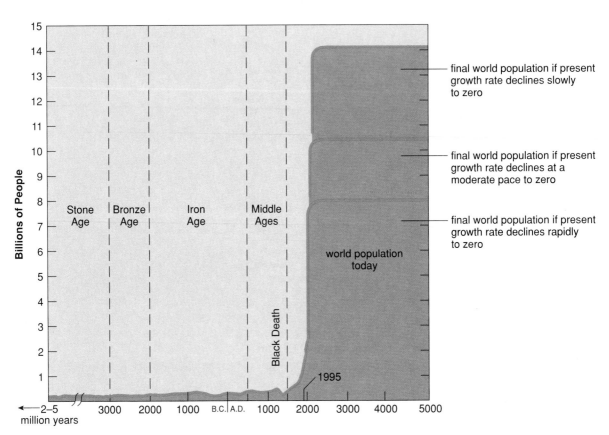

Growth Curve for Human Population
Figure 35.1

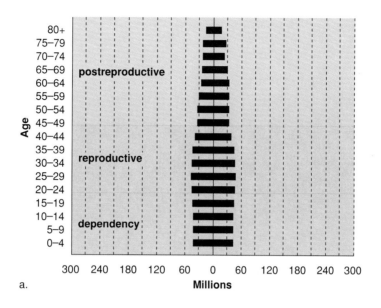

a.

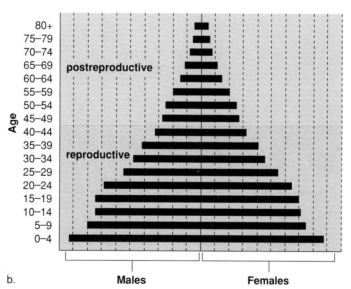

b.

Age-Structure Diagrams for MDCs and LDCs, 1989
Figure 35.4

213

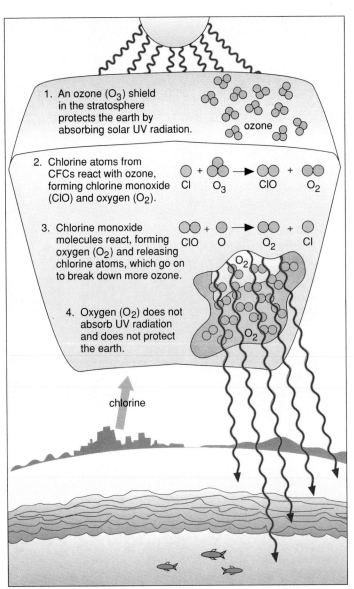

Ozone Hole
Figure 35.7

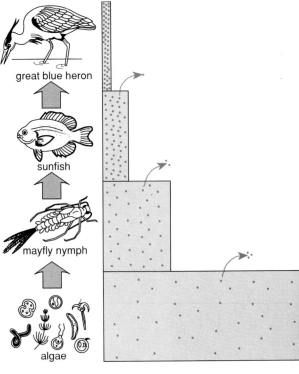

great blue heron

sunfish

mayfly nymph

algae

Biological Magnification
Figure 35.12

CREDITS

Line Art

Fig.15.9 Source: From John W. Hole, Jr., *Human Anatomy & Physiology*, 5th edition. Copyright © 1990 Times Mirror Higher Education Group, Inc., Dubuque, Iowa. All Rights Reserved. Reprinted by permission

Fig.16.2 Source: From Kent M. Van De Graaff and Stuart Ira Fox, *Concepts of Human Anatomy & Physiology*, 3rd edition. Copyright © 1992 Times Mirror Higher Education Group, Inc., Dubuque, Iowa. All Rights Reserved. Reprinted by permission.

Fig.21.1 Source: From John W. Hole, Jr., *Human Anatomy & Physiology*, 6th edition. Copyright © 1993 Times Mirror Higher Education Group, Inc., Dubuque, Iowa. All Rights Reserved. Reprinted by permission.

Fig. 21.5 Source: From John W. Hole, Jr., *Human Anatomy & Physiology*, 6th edition. Copyright © 1993 Times Mirror Higher Education Group, Inc., Dubuque, Iowa. All Rights Reserved. Reprinted by permission.

Figure 35.4 Source: Data from World Population Profile: 1989, WP-89.

Photo credits

Fig.2.19 Photo Credit B: © Dr. Jeremy Burgess/SPL/Photo Researchers, Inc.

Fig. 2.20 Photo Credit B: © Don Fawcett/Photo Researchers, Inc.

Fig. 2.21 Photo Credit B: © Ulrike Welsch/Photo Researchers, Inc

Fig. 3.2 Photo Credit A: © Richard Rodewald/Biological Photo Service

Fig.3.3 Photo Credit A: © W.P. Wergin & E.A. Newcomb/Biological Photo Service

Fig.3.5 Photo Credit A: © W. Rosenberg/Biological Photo Service

Fig. 3.7 Photo Credit A: Courtesy of Dr. Keith Porter

Fig.3.8 Photo Credit A: Courtesy of Herbert W. Israel, Cornell University

Fig.5.1 Photo Credit A1: © Ed Reschke

Fig. 5.1 Photo Credit B1: © Michael Abbey/Photo Researchers, Inc.

Fig.5.5 Photo Credit A1, B1: © Michael Abbey/Photo Researchers, Inc.

Fig. 5.6 Photo Credit A: © Michael Abbey/Photo Researchers, Inc.

Fig.5.8 Photo Credit A1, A2, A3, A4,: Courtesy of Dr. Andrew Bajer

Fig. 7.5 Photo Credit A: Courtesy of Dr. Keith Porter

Fig. 9.7 Photo Credit B: © Carolina Biological Supply/Phototake

Fig. 9.15 Photo Credit B: © Dr. Jeremy Burgess/SPL/Photo Researchers, Inc.

Fig. 9.16 Photo Credit A1: © Renee Lynn/Photo Researchers, Inc.

Fig. 9.16 Photo Credit B1: © William E. Ferguson

Fig. 9.16 Photo Credit C1, D1: © Times Mirror Higher Education Group, Inc./Carolyn verson, photographer.

Fig. 12.6 Photo Credit D: © Manfred Kage/Peter Arnold, Inc.

Fig. 13.11 Photo Credit A: © Lennart Nilsson, *Behold Man,* Little, Brown and Company, Boston

Fig. 14.8 Photo Credit A: © Boehringer Ingelheim International/photo Lennart Nilsson

Fig. 18.1 Photo Credit B,C: © Ed Reschke

Fig. 18.6 Photo Credit A,B,D,E: © Ed Reschke

Fig. 18.6 Photo Credit F: © Ed Reschke/Peter Arnold, Inc

Fig. 18.9 Photo Credit B: Courtesy of H. E. Huxley

Fig.18.11 Photo Credit B: From "Behold Man", Little, Brown, and Co, Boston. Photo by Lennart Nilsson

Fig. 19.10 Photo Credit A: © Lennart Nilsson,

Fig. 19.16 Photo Credit B: © Prof. P. Motta, Dept. of Anatomy, University "La Sapienza", Rome/SPL/Photo Researchers, Inc.

Fig. 21.3 Photo Credit D: © Biophoto Associates/Photo Researchers, Inc.

Fig. 21.7 Photo Credit B: © Ed Reschke/Peter Arnold, Inc.

Fig. 25.1 Photo Credit A: © Lee Simon/Photo Researchers, Inc

Fig. 27.15 Photo Credit B: © Bob Evans/Peter Arnold, Inc.

Fig. 28.11 Photo Credit A: © R. Knauft/Photo Researchers, Inc.

Fig. 28.12 Photo Credit B: © M. I. Walker/Science Source/Photo Researchers, Inc.

Fig. 28.13 Photo Credit B: © William E. Ferguson

Fig. 28.18 Photo Credit B: © Manfred Kage/Peter Arnold, Inc

Fig. 28.19 Photo Credit B: © Carolina Biological Supply/Phototake

Fig. 28.23 Photo Credit B: © David M. Phillips/Visuals Unlimited

Fig. 28.24a Photo Credit A1: © Walter H. Hodge/Peter Arnold, Inc.

Fig. 28.25 Photo Credit A2: © Biophoto Associates

Fig. 28.25 Photo Credit B: © Glenn Oliver/Visuals Unlimited

Fig. 28.27 Photo Credit B: © Stephen Krasemann/Peter Arnold, Inc.

Fig. 29.4 Photo Credit B: © John Gerlach/Visuals Unlimited

Fig. 29.10 Photo Credit B: © Matt Meadows/Peter Arnold, Inc.

Fig. 29.12 Photo Credit B: © Carolina Biological Supply/Phototake

Fig. 30.6 Photo Credit A: © Carolina Biological Supply/Phototake

Fig.30.9 Photo Credit A: © Arthur Siegelman/Visuals unlimited

Fig. 30.9 Photo Credit C: © James Solliday/Biological Photo Service